Dad...

Social Movements and State Power

¡ Hasta la victoria siempre !
Con todo cariño,
Linda y Dave

Social Movements and State Power

Argentina, Brazil, Bolivia, Ecuador

James Petras and Henry Veltmeyer

Pluto Press
LONDON • ANN ARBOR, MI

First published 2005 by Pluto Press
345 Archway Road, London N6 5AA
and 839 Greene Street, Ann Arbor, MI 48106

www.plutobooks.com

British Library Cataloguing in Publication Data
A catalogue record for this book is available from the British Library

ISBN   0 7453 2423 1 hardback
ISBN   0 7453 2422 3 paperback

Library of Congress Cataloging in Publication Data
Petras, James F., 1937–
  Social movements and state power : Argentina, Brazil, Bolivia, Ecuador /
James Petras and Henry Veltmeyer.
    p. cm.
  Includes bibliographical references.
  ISBN 0–7453–2423–1 (hb) — ISBN 0–7453–2422–3 (pb)
  1. Political participation—Latin America—Case studies. 2. Social
movements—Latin America—Case studies. 3. Argentina—Politics and
government—2002– 4. Brazil—Politics and government—1985– 5. Bolivia–
–Politics and government—1982– 6. Ecuador—Politics and government—
1984– I. Veltmeyer, Henry. II. Title.
  JL966.P375 2005
  322.4'098—dc22
                                                        2005005107

10   9   8   7   6   5   4   3   2   1

Designed and produced for Pluto Press by
Chase Publishing Services Ltd, Fortescue, Sidmouth, EX10 9QG, England
Typeset from disk by Stanford DTP Services, Northampton, England
Printed and bound in Canada by Transcontinental Printing

Contents

Preface

Chavez and the Referendum: Myths and Realities

Between rightwing frustration and leftwing euphoria, little has been written about the complex and contradictory reality of Venezuelan politics and the specificities of President Chavez' policies. Even less discussion has focused on the division between ideological Washington and pragmatic Wall Street, between the politics of confrontation and conciliation, and the convergences and divergences between Venezuela and the rest of Latin America. Both the right and left have substituted myths about the Chavez government rather than confronting realities.

RIGHTWING MYTHS

Myth 1—Chavez is an unpopular president whom the rightwing opposition is capable of defeating in the referendum. In reality, the rightwing and its backers in Washington miscalculated on several counts. First, the weakest moment of the Chavez government was immediately after the Petroléos de Venezuela (PDVSA) executive lockout (December 2002–February 2003), when oil prices were much lower, the economy was devastated, government social welfare programs were under-funded and grass-roots political organizations were weak. By the time the referendum took place (August 2004), one and a half years later, socioeconomic and political conditions had dramatically changed. The economy was growing by 12 per cent, oil prices were at record highs, social welfare expenditures were increasing and their social impact was highly visible and widespread, and the mass social organizations were deeply embedded in populous neighborhoods throughout the country. Clearly, the initiative had passed from the right to the left, but both the US and its opposition collaborators were blind to the realities. Having lost control of the state petroleum industry and allocation of funds via the failed lockout in early 2003, and having lost influence in the military after the failed coup of April 2002, the opposition possessed few resources to limit the government's referendum campaign and no leverage in launching a post-election 'civic–military' coup.

Myth 2—According to rightwing analysts the referendum on the opposition's demand that Chavez be recalled from office was based on the issue of Chavez' 'popularity,' 'personality,' charisma, and 'autocratic' style. In reality, the referendum was based on class/race divisions. Non-opposition trade union leaders indicated that over 85 per cent of the working class and working poor voted for Chavez, while preliminary reports on voting in affluent neighborhoods and circumscriptions showed just the reverse: over 80 per cent voted for the referendum. A similar process or class/race polarization was evident in the extraordinary turnout and vote among poor Afro-Venezuelans: the higher the turnout, the higher the vote for Chavez, as an unprecedented 71 per cent of the electorate voted. Clearly, Chavez was successful in linking social welfare programs and class allegiances to electoral behavior.

Myth 3—Among both the right and left there is a belief that the mass media control mass voting behavior, limit political agendas, and necessarily lead to the victory of the right and the domestication of the left. In Venezuela the right controlled 90 per cent of the major television networks and print media and most of the major radio stations, yet the referendum was crushed by an 18 per cent margin (59 per cent to 41 per cent). The results of the referendum demonstrate that powerful grass-roots organizations built around successful struggles for social reforms can create a mass political and social consciousness which can easily reject media manipulation. Elite optimism in their 'structural power'—money, media monopoly, and backing by Washington—blinded them to the fact that conscious collective organization can be a formidable counterweight to elite resources. Likewise, the referendum results refute the argument of the center-left that they lose elections because of the mass media. The center-left justify their embrace of neoliberalism as a means to 'neutralize' the mass media during elections. They refuse to recognize that elections can be won despite mass media opposition if previous mass struggle and organization create mass social consciousness.

Myth 4—According to many leftist journalists, Chavez' victory reflected a new wave of popular nationalist politics in Latin America. Evidence to the contrary is abundant. Brazil under Lula has sold oil exploration rights to US and European multinational corporations, provides a contingent of 1,500 troops (along with Argentina, Chile, etc.) to Haiti to stabilize Washington's puppet regime there imposed through the kidnapping of President-elect Aristide. Likewise in the other Andean countries (Ecuador, Peru, Bolivia, and Colombia) the elected regimes propose privatizing petroleum companies, support ALCA and Plan Colombia, and pay their foreign debts. The Broad Front in Uruguay promises to follow Brazil's neoliberal policies. While Chavez promotes the regional trading bloc, Mercosur, the two major members—Brazil and Argentina—are increasing their trade relations outside the region. In effect, there is a bloc of neoliberal regimes arrayed against Chavez' anti-imperialist policies and mass social movements. To the extent that Chavez remains true to his independent foreign policy, his principal allies are the mass social movements and Cuba.

Myth 5—The defeat of the referendum was a major tactical defeat of US imperialism and its local vassals. But a defeat of imperialism does not necessarily mean or lead to a revolutionary transformation, as post-Chavez post-election appeals to Washington and big business demonstrate. More indicative of Chavez' politics is the forthcoming $5 billion investment agreements with Texaco-Mobil and Exxon to exploit the Orinoco gas and oil fields. The euphoria of the left blinds them to the pendulum shifts in Chavez' discourse and the heterodox social welfare—neoliberal economic politics he has consistently practiced.

President Chavez' policy has always been a careful balancing act between rejecting vassalage to the US and local oligarchic rentiers on the one hand, and trying to harness a coalition of foreign and national investors, and urban and rural poor, to a program of welfare capitalism on the other. He is closer to Franklin D. Roosevelt's New Deal than to Fidel Castro's socialist revolution. In the aftermath of the three political crises—the failed civil–military coup, the debacle of the corporate executives' lock-out, and the defeat of the referendum— Chavez offered to dialogue and reach a consensus with the media barons, big business plutocrats, and the US government, on the basis of the existing property relations, media ownership, and expanded relations with Washington.

Chavez' commitment to centrist-reformist policies explains why he did not prosecute owners of the mass media who had openly called for the violent overthrow of his government and also why he took no action against the association of the business leaders (FEDECAMARAS) who had incited military rebellion and violent attacks on the constitutional order. In Europe, North America, and many other regions, democratically elected governments would have arrested and prosecuted these elites for acts of violent subversion. President Chavez has constantly reiterated that their property, privileges, and wealth are not in question. Moreover, the fact that these elites have been able to engage in three unconstitutional attempts to overthrow the regime and still retain their class positions strongly suggests that President Chavez still envisages their playing an important role in his proposed development based on private–public partnership and social welfare spending. After five years in government and three major 'class confrontations,' it is evident that at least at the level of the government, there has been no rupture in property or class relations and no break with foreign creditors, investors, or oil clients. Within the fiscal framework of foreign debt payments, subsidies to private exporters, and low-interest loans to industrialists, the government has increased the allocation of state spending for social programs in health, education, housing, micro-enterprises, and agrarian reform. The Venezuelan government can maintain this balance between big business and the poor because of the high prices and revenue from petroleum exports. Like President Roosevelt, Chavez' positive social welfare programs attract millions of low-income voters, but do not affect money income levels or create large-scale employment projects. Unemployment is still in the region of 20 per cent and poverty levels remain at over 50 per cent. Comprehensive social spending has positively affected the social lives of the poor but has not improved their class position. Chavez is both confrontational and radical when his rulership is threatened, and conciliatory and moderate when he successfully overcomes the challenge.

Myth 6—The left and right have failed to recognize a divergence of tactics between an ideological Washington and a pragmatic Wall Street. The US political class (both Republican and Democrats, the presidency and Congress) have been actively threatening, intervening, and supporting destructive lock-outs, violent coups, and a fraudulent referendum to oust Chavez. In contrast, the major American and European oil companies and banks have been engaged in stable, sustained, and profitable economic relations with the Chavez government. Foreign creditors have received prompt and punctual payments of billions of dollars and have not spoken or acted in a fashion to disrupt these lucrative transactions. Major American multinational oil companies project investing between $5 billion and $20 billion in new exploration and exploitation. No doubt these MNCs would have liked the coup to succeed in order to monopolize all Venezuelan oil revenue, but perceiving the failures of Washington they are content to share some of the oil wealth with the Chavez regime. The tactical divergences between Washington and Wall Street are likely to narrow as the Venezuelan government moves into the new conciliatory phase toward FEDECAMARAS and Washington. Given Washington's defeat in the referendum, and the big oil deals with key American multinationals, it is likely that Washington will seek a temporary 'truce' until new, more favorable circumstances emerge. It will be interesting to see how this possible 'truce' will affect Venezuela's critical foreign policy.

Myth 7—The main thrust of the current phase of the Chavez revolution is a moral crusade against government corruption and a highly politicized judicial system tightly aligned with the discredited political opposition. For many on the left, the radical content of the 'No' vote campaign was rooted in the proliferation of community-based mass organizations, the mobilization of trade union assemblies, and the decentralized democratic process of voter involvement based on promises of future consequential social changes in terms of jobs, income, and popular political power.

Moralization campaigns (anti-corruption) are commonly associated with middle-class politics designed to create 'national unity' and usually weaken class solidarity. The left's belief that the mass organizations mobilized for the referendum will necessarily become a basis for a 'new popular democracy' has little basis in the recent past (similar mobilizations took place prior to the failed coup and during the corporate bosses' lock-out in mid-April 2002).[1] Nor do government-sponsored moralization campaigns attract much interest among the poor in Venezuela or elsewhere. Moreover, the focus of the *Chavista* political leaders is on the forthcoming elections for parliament, not in creating alternative sources of governance. The left's facile projection of popular mobilization in the post-referendum period creates a political mythology that fails to recognize the internal contradictions of the political process in Venezuela.

CONCLUSION

The massive popular victory of the 'No' vote in the Venezuelan referendum gave hope and inspiration to hundreds of millions in Latin

America and elsewhere that US-backed oligarchies can be defeated at the ballot box. The fact that the favorable voting outcome was recognized by the Organization of American States (OAS), President Jimmy Carter, and Washington is a tribute to President Chavez' strategic changes in the military, guaranteeing the honoring of the constitutional outcome. At a deeper level of analysis, the conceptions and perceptions of the major antagonists among the right and the left, however, are open to criticism: the right for underestimating the political and institutional support for Chavez in the current conjuncture; the left for projecting an overly radical vision on the direction of politics in the post-referendum period. From a 'realist' position, we can conclude that the Chavez government will proceed with his 'New Deal' social welfare programs while deepening ties with major foreign and domestic investors. His ability to balance classes, leaning in one direction or the other, will depend on the continued flow of high returns from oil revenues. If oil prices drop, hard choices will have to be made—class choices.

Introduction

The 2003 electoral victory of Lucio Gutiérrez in Ecuador was greeted with the same sense of optimism and expectation of a new direction and alternative politics that greeted the election to the presidency of Ignacio ('Lula') da Silva in Brazil and has surrounded Hugo Chavez' declaration of the Bolivian revolution. As for Gutiérrez's ascendancy to state power, it was viewed as a major political advance for the country's indigenous nationalities in its 500 year-long struggle for freedom (from oppression and exploitation) and democracy (participation in the direction of national policy and the country's political affairs). In Bolivia, Evo Morales, the leader of the *cocaleros*, an organization of coca-producing peasants, as the candidate for the *Movement Towards Socialism*, came within an electoral whisker of achieving state power in 2002. This development awakened hopes on the left of a possible new dawn in Latin American politics. Even the ascendancy of Nestor Kirchner to the presidency in Argentina, after an economic and political crisis of historic proportions, encouraged the same hopes and expectations for a fundamental change of direction in national policy—at least as regards the neoliberal model of free-market capitalist development and globalization. In each case, and collectively, political developments in some of the region's largest and most important countries have been widely viewed on the left as growing evidence of both the demise of neoliberalism and the power of the US to shape economic policy in the region, as well as representing a new wave of progressive regimes, constituting an anti-US axis in foreign policy, oriented toward an alternate popular form of national development.

Notwithstanding the title of a recent Italian film on the elections in Ecuador (*How George W. Bush Won the Elections*), the electoral victory of Gutiérrez was widely interpreted on the left as a setback to the efforts of the US government to dominate economic and political developments in the region—to reassert its hegemony and what some neoconservative advisors to the administration see as a project of imperial domination—a 'new imperialism' that is not afraid to speak its name or assert state power.

Several years after or into these political developments in Argentina, Bolivia, Brazil, and Ecuador, and elsewhere in the region, a number of fundamental questions have arisen, questions that are addressed in this volume. They include the following: How realistic were and are the expectations of the left of progressive change in the region? What is the political significance of the road or rise to power of Lula in Brazil, Kirchner in Argentina, Gutiérrez in Ecuador, and Morales in Bolivia? What are the dynamics of electoral politics relative to the struggle for state power? How do these dynamics relate to the strategy and tactics of mass mobilization preferred by the social movements? What are the advantages and limits of both paths to state power? What are the dynamics of the relationship between the social movements and the state? What are the theoretical and political implications of this relationship?

Some of these questions have already been raised in regard to Da Silva (Lula)'s ascendancy to state power in Brazil. In this case the left's expectations and illusions have already been shattered. The Gutiérrez regime? His election was also widely viewed as a victory for the left and a blow to the US and its agenda for Latin America. Kirchner? His regime has raised a number of still unanswered questions about the politics of state power in the new regional and global context—and the prospects for politically-driven social change. And Morales? His leadership raises serious questions about the most efficacious way in which the organized popular movement might achieve state power and advance the struggle for substantive social change.

This volume examines these and other such questions within the framework of the theoretical issues that surround the relationship between the state and social movements. One of these issues has to do with the question of state power and the different paths toward achieving it—electoral politics, the path preferred by the 'political class' because it is predicated on limited political reforms to the existing system, or mass mobilization of the forces of resistance and opposition. This is the path taken by most social movements. It is oriented toward more fundamental or radical change in the existing system. In addition, there are two further conceptions of political practice and power. One is associated with a postmodernist perspective on a new form of 'politics' and the emergence in theory (that is, academic discourse) of the 'new social movements.'

In this perspective, the way to bring about social change is not so much through political action in the struggle for state power as through social action involving the construction of a 'no-power'— in social relations of coexistence, solidarity, and collective action. Another approach toward social change is associated with the nongovernmental organizations (NGOs) involved in the process of international development. The ostensible aim of this strategic 'project' is to form partnerships with governments and overseas development assistance organizations in promoting and improving the lives of the poor—to bring about conditions that will sustain their livelihoods and alleviate their poverty, and to do so not through a change in the structure of economic and political power but through an empowerment of the poor. This approach seeks to build on the social capital of the poor themselves, seeking thereby to improve their lives within the local spaces available within the power structure. It is predicated on partnership with likeminded organizations in a shared project (the alleviation of poverty, sustainable livelihoods). Rather than directly confronting this structure in an effort to change the existing distribution of power, the aim, in effect, is to empower the poor without disempowering the rich.

Under conditions found across Latin America, and experienced in different ways in virtually every country, each of these forms of social change and associated conceptions of power have their adherents. However, in the current conjuncture, exemplified in the country case studies included in this volume, the proponents of 'a new way of doing politics,' i.e., the no-power, alternative development approach to social change advocated by Antonio Negri, John Holloway, and others (Dieterich, for example), are totally irrelevant, unable to explain or inform the politics of widespread resistance to capitalist development and imperialism in neoliberal form. All of the social movements, from the least to the most consequential and dynamic, are engaged in a struggle for state power. Nevertheless, in regards to the political dynamics of this struggle there is a series of critical issues that remain to be settled in both theory and practice. This is our main purpose in writing this book, and is the central object of analysis in each case study.

The volume begins with an analysis of the political dynamics involved in diverse efforts to create a 'new' world economic and political order. The economic conditions of this 'new world order' were created on the basis of a 'new economic model,' which provided the underpinnings of a process of structural adjustment, globalization, and neoliberal capitalist development. This dynamic of the economic reform process has been studied in some depth by various scholars, including the contributors to Veltmeyer and Petras (1997; 2000) and Petras and Veltmeyer (2001; 2003). However, the political dimensions of this process—the politics of adjustment (to secure the necessary conditions of governability or 'good governance')—require a close look and more study. This chapter focuses on diverse and widespread efforts in this regard, efforts that include a proposed marriage between capitalism and democracy—to democratize the state and its relation to 'civil society'; and to strengthen civil society and engage stakeholder organizations in a project of 'good governance.' The chapter examines critically the agenda behind this project. It also focuses on the role of NGOs, a key sector of 'civil society,' as a Trojan horse for global neoliberalism, as an agency for helping the guardians of the New World Order secure the political conditions needed to pursue their neoliberal policy agenda.

Chapter 2 takes up the case of Argentina. Between December 19 and 21, 2001 a massive popular rebellion overthrew the incumbent president amidst the fiercest street battles and highest casualties in recent Argentine history. Major demonstrations and street blockades took place throughout the country, in an unprecedented alliance between the mass of unemployed workers and a substantial sector of the middle class defrauded of their savings. In quick succession three congressional aspirants who sought to replace President De la Rua were forced to resign. From December 2001 to July 2002, the popular movements were a force in the streets and a visible presence in all provinces, blocking highways as well as the major thoroughfares of Buenos Aires and provincial capitals. It is estimated that up to four million persons (out of a potentially active population of less than 30 million) took part in the demonstrations. Numerous writers spoke of a 'pre-revolutionary situation'; they wrote of 'dual power' between the '*piqueteros*', neighborhood assemblies, and the 'occupied factories' on the one hand, and the existing state apparatus, on the other. All of the divisions and agents of the state apparatus (the judiciary, the police, the armed forces) as well as the traditional parties, politicians, and Congress lost legitimacy in the eyes of the majority of Argentinians in the events leading up to and immediately after the uprising of December 2001.

This chapter takes the events of December 19–21 as the basis of an analysis of subsequent developments in the relationship of the state to the popular movement. At the center of this relationship is Nestor Kirchner, a politician of the *Justicialista* Party, which was responsible for bringing Argentina to the brink, but who, as the latest occupant of the 'Pink House' (the executive branch of the government), has managed to restore a measure of economic and political order. The political dynamics of the Kirchner regime since December 2001 allow the authors to reflect on the theoretical and political issues involved in the class struggle for political power. The chapter provides a detailed analysis of the dynamics of this struggle. It is concluded that in the current context Kirchner has neither the economic resources nor the ideology to sustain his delicate balancing act. At the same time, the mass movement has learnt and is continuing to learn important lessons about the best way to advance in its protracted struggle for political power as a means of bringing about substantive social change.

Chapter 3 examines the political dynamics of the regime established by Da Silva in Brazil. The stakes were high. No regime in the region has raised so high the expectations on the left for a fundamental realignment of state power and the possible emergence of an anti-US/imperialism bloc. And no regime has as thoroughly dashed these hopes and expectations. The authors explore in some detail and analyze the complex dynamics of the forces at play in Lula's politics and policies over the past two years. The dynamics of these forces revolve around the relationship of Lula's neoliberal policy regime to foreign capital and Washington on the one hand, and the dominant class and the popular movement headed by the Rural Landless Workers on the other. At issue in this relationship is political power—recourse to, and use of, diverse political instruments to advance the agenda and economic interests of a small but dominant social class and to frustrate the hopes and expectations of the working class and the popular movement. The result, as discussed in Chapter 3, is a deepening of the social crisis and 'opulence in the midst of misery.'

In the context of this growing social crisis for the struggle for political and state power the issue resolves into the question of the limits of electoral politics and the dynamics of social movements. The authors draw a number of conclusions from these limits and dynamics in regard to the prospects for fundamental social change in Brazil.

Chapter 4 turns to Bolivia, where the Movement Towards Socialism (MAS), led by the leader of a movement of 30,000 coca-producing indigenous peasants from the eastern lowlands, almost took state power in the elections of 2002. The presidential candidate of the MAS came within a few votes (1.4 per cent of the total cast) of winning the election and thereby gaining access to, and nominal control over, the state apparatus. This event has enhanced the credibility of the parliamentary or electoral road to state power as opposed to the strategy of mass mobilization which is preferred by the social movements. But an analysis of the dynamics of electoral politics and the mass mobilizations of the social movements leads the authors to a different conclusion.

The dynamics involved in the struggle for political power in Bolivia can be understood only within the context of the politics and economics of adjustment made at the level of the state. In this context the chapter shows Bolivia to be a crucible for diverse experiments with policies that might serve to establish political order for a neoliberal policy regime. A review of the politics and economics of Bolivia's adjustment to the New World Order provides the authors with a framework for a class analysis of the dynamics involved in the state–social movement relation.

Such an analysis, used in each of the four case studies in this volume, has three crucial steps: first, an analysis of the forces generated by the structure of the system and shaped by its objectively given conditions; second, an analysis of the social composition of the forces generated under these conditions (at this level issues of ethnicity and gender arise, as does the issue of characterizing in social terms the base of the social movement, but an analysis of these issues is framed by a class analysis); and third, a political–dynamic analysis of the forces of resistance and opposition generated under these conditions.

In Chapter 5 the authors' spotlight is turned on Ecuador, which has witnessed the rise of one of the region's most dynamic social movements, a movement with its social and organizational base in the country's indigenous communities and the historic struggle of these communities for land, democracy, and social justice as well as respect for indigenous ways of life and the construction of a multi-ethnic state in which the indigenous nationalities can move toward or achieve a measure of autonomy, self-determination, full participation, and equity.

The dynamics of this political struggle are complex and, as elsewhere, derive from the relationship between, on the one hand, the state and its neoliberal policy agenda, and, on the other, the popular movement and its agenda of fundamental social change. The political dynamics of this relationship are explored within the context of close to two decades of neoliberal policies and organized resistance against them. Important moments in this struggle include several indigenous uprisings and the building of a popular movement that has brought together communities and diverse organizations of indigenous people with the highland peasantry in their land struggle and the working class and their struggle. The chapter highlights Gutiérrez's election in 2002, with the support of CONAIE, the most important and dynamic part of the indigenous movement. The chapter then proceeds to analyze the political dynamics associated with the subsequent regime and its evolving relationship with both the indigenous movement and the broader popular movement. The dynamics of this relationship are analyzed in terms of the options available to the popular movement in its quest for political power, i.e. the capacity to help shape policies related to the 'authoritative allocation of society's productive resources.' These options include: (1) participation in a system of electoral politics governed by the rules of the 'political class'; (2) participation in the politics of alternative development in the form of 'micro-projects' for local development; and (3) the mass mobilization of the forcers of resistance and political opposition.

The concluding chapter draws out the theoretical—and political—implications of the country case-studies presented in this volume and brings together the various threads of our analyses. The result is not a general theory but a series of generalizations relating to the political dynamics of an ongoing class struggle for political power in Latin America. These generalizations allow the authors to reflect on, if not explain, the complex dynamics of a process that is shown to be neither predetermined nor open-ended. As we see it, and argue in this chapter, substantive or antisystemic social change requires a closer look at the issues addressed in this volume and an active, informed engagement with these issues in practice as well as thought. In other words, what is required is an advance of the popular movement toward social revolution. The authors hope to have made a modest contribution toward this project.

1
Bad Government, Good Governance: Civil Society versus Social Movements

Few terms have achieved such wide currency as 'globalization' to describe the epoch-defining changes that have characterized the past two decades and to prescribe a policy agenda of privatization, liberalization, and deregulation. As description and prescription, globalization is associated with neoliberal policies of structural adjustment that are designed to create a worldwide capitalist economy organized so as to release the forces of 'freedom, democracy, and private enterprise'—to quote from the National Security Report submitted to the US Congress by George W. Bush in September 2002. However, the promoters and guardians of this New World Order have not had an easy time of it. For one thing, the social inequalities generated in the process have not only spawned growing levels and diverse forms of discontent and social conflict, but the resulting forces of resistance against global capitalist development have been leveled against the system, undermining and weakening the neoliberal regimes committed to policies of adjustment and globalization. Under these conditions the international organizations behind this project have had to confront a serious political issue of 'governability'—i.e., the ungovernability of the forces freed from the constraints of state regulation (World Bank, 1994; Bardhan, 1997; Kaufmann, Kraay and Zoido-Lobatón, 1999).[1]

'Governance' in this situation is defined as replacing the mechanisms of political control hitherto associated with the nation state. Within the neoliberal model the state is viewed in two ways. On the one hand, it is viewed as Adam Smith saw it—as a predatory device and, in the language of the 'new political economy,' susceptible to rentierism and corruption. On the other, the 'state' (the government, to be more precise) is viewed as an inefficient means of allocating society's productive resources for the social distribution of national income. In this connection, within the parameters of the old and now defunct economic model in place since the 1950s, the government, in its policies of strategic nationalization, protectionism vis-à-vis domestic industry and local enterprise, and market regulations, distorted the normal working of the market, leading to withdrawal of capital from the production process and generating thereby widespread problems of informalization, poverty, and unemployment as well as fiscal imbalance. Within these optics, the state has been subjected to pressures for institutional and policy reform—in the direction of macroeconomic equilibrium (balanced budgets and accounts) and structural adjustment and administrative decentralization; to surrender thereby its capacity for resource allocation and reduce its economic role vis-à-vis responsibilities for development and social programming. A more strictly political dimension to these reforms was democratization, not so much in a return to the rule of law and electoral politics as a change in the relation of the state to civil society. This, we argue, is the crux of the governability/governance issue.

The problem—a problem, that is, for the promoters of globalization— is that neoliberalism is economically dysfunctional, in social terms profoundly exclusionary, and politically unsustainable, generating destabilizing forces of resistance in the form of antisystemic social movements. It is precisely as a means of dealing with this problem that the international organizations for development and finance have turned so decisively toward democratization and civil society, contracting nonprofit voluntary associations (NGOs) and converting them into their agents as 'strategic partners.' The agenda in this strategy is to enlist the help of these NGOs in dousing the fire of revolutionary ferment in the countryside—to provide the rural poor and the popular sector of society with an alternative to the social movements and their radical antisystemic politics.

This chapter explores diverse dynamics of this political process. It means that the social movements, which were formed as means of resisting the neoliberal agenda of globalization and free-market development, are beset by forces designed to demobilize them— to divert the struggle for state power in one or more directions toward electoral politics, reformist social organizations, or local development.

CIVIL SOCIETY, DEVELOPMENT, AND DEMOCRACY

Globalization is one of several ideas advanced in the lexicon of the new 'global economy.' The marketing (by advocates of a New World Order) of 'globalization' as a policy prescription of deregulation, liberalization, and privatization was accompanied by the resurrection of a term used by the rationalist humanists of the eighteenth-century Enlightenment (the 'Age of Reason') to distinguish a sphere independent from the state, namely 'civil society.' In the context of a neoconservative attack on the welfare/developmentalist state the idea of 'civil society' achieved prominence in political and developmental discourse, particularly in connection with successive waves of democratization, beginning in Latin America and Eastern Europe, and spreading across the developing world. In this context 'civil society' was seen as an agent for limiting authoritarian government, strengthening popular empowerment, reducing the socially atomizing and unsettling effects of market forces, enforcing political accountability, and improving the quality and inclusiveness of governance. Reconsideration of the limits of state action also led to an increased awareness of the potential role of civic organizations in the provision of public goods and social services, either separately or in a 'synergistic' relationship with state institutions. In this context, the idea of 'civil society,' like that of 'globalization,' was converted into a discursive weapon and ideological tool in service of advancing the neoliberal agenda.

The academic discourse on civil society, however, has moved beyond this agenda and can now be put into three ideological categories—conservative, liberal, and radical. On this ideological spectrum liberals generally see civil society as a countervailing force against an unresponsive and corrupt state and exploitative corporations that ignore environmental issues and human rights abuses (Kamat, 2003). Conservatives, on the other hand, see in civil society the beneficial effects of globalization for the development of democracy and economic progress—for advancing the idea of freedom in its historic march against its enemies (Chan, 2001). As for those scholars that share a belief in the need for radical change, civil society is seen as a repository of the forces of resistance and opposition, forces that can be mobilized into a counter-hegemonic bloc or a global anti-mobilization movement (Morton, 2004).

In effect, academic discourse in its diverse ideological currents appears to converge on civil society, viewing it as an agent for change in one form or another. The growth and strengthening of 'civil society' (nongovernmental, social and civic organizations) in the 1980s and 1990s are offered as proof of its capacity for autonomous development and the virtues of 'democracy,' a state that is subject to powerful democratizing tendencies and forces that favor democratic renewal. In this process of democratic renewal (or 're-democratization', as it is referred to in the literature) NGOs are assigned a leading role as front-line agents of a participatory and democratic form of development and politics, to convince the rural poor thereby of the virtues of community-based local development and the need to reject the confrontational politics of the social movements.

In the 1980s there was a veritable explosion of NGOs, many of which were formed in the wake of a retreating state. It is estimated that the vast majority of the 37,000 or so NGOs operating today in diverse developing countries were formed in the 1980s or the 1990s. As noted above, the NGOs in this historic context were contracted by international organizations—and the governments engaged in the international development project—to spread the gospel of the free market and democracy and to speak of the virtues of social democratic 'civic' organization and action within the local spaces available within the national power structure. Despite the serious reservations of many governments in the developing world (because of their 'politics'), the NGOs in this project were viewed as vastly preferable to the social movements that were generally oriented toward collective action against the power structure, seeking to change this structure rather than search for accommodation with it. In this political context the NGOs are enlisted by overseas development assistance agencies (ODAs) and governments as partners in the process of 'sustainable human development' and 'good [democratic] governance'—as watchdogs of state deviancy, to ensure its transparency (inhibit or prevent corruption and rentierism), and as participants in the formulation of public policy. The institutional framework for this 'participatory' form of development and politics (and governance rather than government) would be established by the decentralization of decision-making capacity and associated responsibilities from the national to the local level and the institution of 'good governance,' that is, a democratic regime in which the responsibility for human security and political order is not restricted to the government and other institutions of the state but is widely shared by different civil society organizations (World Bank, 1994; BID, 1996, 2000; UNDP, 1996; OECD, 1997).

The global phenomenon and explosive growth of NGOs reflects a new policy and political consensus that they are *de facto*, and by design, effective agents for democratic change, an important means for instituting an alternative form of development that is initiated from below, and is socially inclusive, equitable, participatory, and sustainable. This consensus view is reinforced by evidence that the NGO channel of overseas development assistance by and large is dedicated to political rather than economic development—to ensure transparency (to inhibit or prevent government corruption and rentierism), promote democracy in the process of change, inculcate relevant values and respect for democratic norms of behavior, and encourage the adoption of 'civil' politics (dialogue, consultations, negotiation) rather than the confrontationalist politics of the social movements.

The leading role of 'civil society organizations' (CSOs) in this regard (political development) foretells a reworking of 'democracy' in ways that coalesce with global capitalism and the neoliberal agenda. Indeed, a well-placed development practitioner in the UK (Wallace, 2003) has wondered aloud (and put in print) whether NGOs in this regard have been used by the international organizations as their stalking horse—and, not to put too fine a point on it, as an agent of global neoliberalism. Global policy forums and institutions, such as the OECD's Development Center, USAID, the World Bank, and the Inter-American Development Bank, as well as the UN's operational agencies such as the UNDP, have all turned to the NGOs as 'forces of democratization' in the 'economic reform process' (Kamat, 2003: 65). In this, Ottaway (2003: vi) argues, they function as agents of 'democratic promotion,' a 'new activity in which the aid agencies and NGOs [originally] embarked [upon] with some trepidation and misgivings,' but that in the early 1990s 'came of age.'

CIVIL SOCIETY AND THE STATE

In the 1990s the perception of NGOs as 'Trojan horses for global neoliberalism' (Wallace, 2003) also 'came of age.' But the effectiveness of NGOs in this regard is not without controversy. Indeed, it has occasioned something of a debate between liberals, in general favorably disposed toward the NGOs, and conservatives, who view them as 'false saviors of international development' (Kamat, 2003).[2] In the same context, radical political economists—and we situate ourselves here—tend to view NGOs as agents (or instruments), either knowingly or often unwittingly, of outside interests. And, in the same context, both economic development and democracy appear as masks for an otherwise hidden agenda, used to impose the policy and institutional framework of the new world order against resistance.

This apparent convergence between the left and the right in a critical assessment of development NGOs points toward several problems involved in the use of the state as an instrument of political power. From a liberal reformist perspective the state needs to be strengthened, but it also needs to be democratized in the service of a more inclusive and participatory approach to policy design and implementation. From a neoliberal (and politically conservative) perspective, however, the state *is* the problem. On the one hand, it is an inefficient means of allocating the productive resources of the system. On the other, as Adam Smith argued, it is a predatory device with a tendency to serve special interests and to capture rents from state-sponsored and regulated economic activities. The state officials, it is added by contemporary advocates of this view, such as the economists at the World Bank, are subject to pressures that more often than not result in their corruption. The solution: a minimalist state, subject to the democratizing pressures of civil society, i.e. groups and organizations able to secure the transparency of the policymaking process.[3]

And what of the state as viewed through the lens of an alternative, more radical political economy? From this perspective the state is an instrument of class rule, a repository of concentrated political power needed to turn the process of national development around—in a socialist direction. In this context, the essence of what is now widely regarded as the politics of the old left—or the old politics of the left—is a struggle over state power. Both leftist political parties and the social movements tend to be oriented in this direction, albeit in a new political context which has seen the emergence of a new perspective and way of doing politics—the politics of 'no-power,' which is to avoid confrontations with the structures of political and economic power and instead to build on the social capital of the poor, to engage in projects of local development.

In the academic world the politics of state power is theoretically constructed in these ways. But what about in the real world? In this context, and with specific reference to developments in Latin America, the main pattern of political development over the past two decades seems to have been a twofold devolution/involution of state power. On the one hand, the policy and institutional framework for political decision-making has been subjected to what has been termed the Washington Consensus, with a corresponding shift of political power (vis-à-vis macroeconomic policy) toward Washington-based 'international' institutions such as the World Bank and the IMF. On the other hand, a democratic 'reform' process has resulted in the institution of the 'rule of law' and the decentralization of government from the center to the local as well as the strengthening of civil society, viz. its capacity to participate in public policymaking.

The latter development is based on various forms of partnerships between international organizations and governments on the one hand, and civil society on the other. And this development was no coincidence. It is based on a conscious strategy pursued by the major representative organizations of global capital and the new world economic order—the imperial brain trust, as Salbuchi (2000), defines these organizations. Among these organizations can be found the World Bank, the regional banks such as the IDB, ODAs such as USAID, the Development Center of the OECD, and operational agencies of the UN system such as the UNDP, ENEP, FAO, WHO. Each of these organizations pursues a partnership strategy with NGOs and other civil society organizations, setting up a division (or 'office') to work with them, officially registering those prepared to cooperate with them in a common agenda of democratic development, poverty alleviation, and environmental protection—an alternative form of participatory, socially inclusive, and 'human' (economic and social) development.

In this context, much of the current academic discourse on the role of NGOs in the economic and political development process focuses on the issue of improving their organizational effectiveness as well as their accountability—as well as their 'autonomy' vis-à-vis governments and the donor organizations (the ODAs). As for the latter, several umbrella organizations within the NGO sector have sought assiduously to ensure greater independence from both donors and the governments that hire 'private voluntary organizations' (PVOs) to execute their projects and programs. But, generally speaking, these efforts have not met with any success. More often than not, as in the case of the US, the major NGOs have not only met with resistance on the part of the donor community but outright efforts to bring NGOs into line. As for USAID, in 2003 the director at the time bluntly informed an assembly of NGOs brought together by Interaction, an umbrella organization of American NGOs, that they would have to do a better job acknowledging their ties to government, as private contractors of public policy, or risk losing funding. Our own research (see also Okonski, 2001) indicates that a substantial number of NGOs in recent years in fact have become increasingly dependent on this funding.

Some go as far as to argue that the presumed role of the NGO is a smokescreen that obscures the workings (and interests) of a powerful state (imperialism), various national elites, and the predations of private capital. Hayden (2002) argues this from a conservative perspective. We, however, argue the same point from a more radical perspective on NGOs as agents of imperialism—private contractors of governments in the North, particularly the US. Governments in the South, in many cases, are only reluctantly and belatedly moving away from a somewhat skeptical, if not hostile, attitude—born of experiences with NGOs as watchdogs of the state, particularly in terms of any propensities toward authoritarianism and corruption, from the perspective of an agenda to promote democracy in its relation to civil society. In a situation of widespread authoritarianism, violation of human rights and other abuses of political power the NGOs throughout the 1980s had no fundamental problem in assuming their intermediary role in the front line of economic and political development. However, in the changed, more democratic context of the 1990s many NGOs began to voice serious concerns that, in effect (by design if not intent), they were advancing the agenda of the donors rather than that of the urban and rural poor, many of whom were not oriented toward alternative development and representative democracy but more substantive social change based on direct action and social movements—that is, popular democracy. In this context the major NGOs redoubled their efforts to secure greater autonomy from donors to be able thereby to respond better to the concerns and priorities of the popular movement. As a result, they tend to find themselves caught between a widespread concern to increase their independence from their sponsors and the efforts of these sponsoring organizations to incorporate them into the development and political process as strategic partners in a common agenda.

NGOs AND THE NEW POLICY AGENDA

In the 1980s organizations of international cooperation for development were fundamentally concerned to convert PVOs into development agencies that could mediate between official, aid-providing agencies and grass-roots communities in the delivery of ODA (Tendler, 1975); and, in the same context, to promote democracy both in the relation of the state to civil society and in the politics of grass-roots organizations—'good governance' in the official parlance (BID, 1996; UNDP, 1996; Blair, 1997; OECD, 1997; Annan, 1998; Mitlin, 1998; Kaufman et al., 1999).

In the late 1980s and early 1990s, however, there occurred a marked shift in practice signaled by a change in discourse—from a 'third sector' discourse privileging NGOs to a civil society discourse that was more inclusive, particularly as regards profit-making enterprises and business associations that make up the 'private sector' (Mitlin, 1998). This shift in discourse coincided with widespread recognition in official circles of the need to reform the structural adjustment program—to give it a social dimension (a new social policy) and the whole process a 'human face' (Cornia et al., 1987; Salop, 1992).

Political discourse in the 1980s reflected the political dynamics of an ideological shift from a state-centered or led development process to a market-led form of development based on the privatization of public enterprise. A 'third sector' discourse in this context represented a concern for an alternative, more participatory form of development and politics predicated on neither the agency of the state ('from above') nor the workings of the market ('from the outside'), but initiated within civil society ('from below'). From the perspective of the ODAs, the international financial institutions (IFIs), and governments, however, this discourse was problematic in various regards. For one thing, it was directed against both the market and the state—against public *and* private enterprise. For another, it worked against the efforts of the ODAs to incorporate the private sector into the development process. The problem was twofold. One was how to overcome widespread antipathy toward profit-making 'private' enterprise—to see it as part of a possible solution rather than as a major problem. The other was to convince the private sector that profits can be made in the process of social development.

As for the second problem it remains a concern even in the twenty-first century, making it difficult for the UN's ongoing efforts to establish its 'global compact' with the private sector (UNDP, 1998; Utting, 2000). As regards the first problem, however, a civil society discourse has proved to be both useful and effective. It has indeed allowed the ODA community to incorporate the private sector into the development project as a strategic partner in the process of economic growth and 'sustainable human development.' The perceived need for this was established by evaluation studies that suggested that NGOs did indeed provide a useful channel for ODAs in regard to political development (promotion of democracy) and capacity-building/strengthening (social capital), but an inefficient means of activating production and employment and providing 'financial services.' In this regard, the conclusion was drawn that what was needed was a new strategy based on the agency of local governments working in partnership with ODAs and NGOs. To this we now turn.

GOOD GOVERNANCE AND ALTERNATIVE DEVELOPMENT

The evolution of CBOs (community-based development organizations) or GROs (grass-roots development organizations) within civil society illustrates the changed environment in which NGOs now operate. For Kamat (2003: 65) it also points toward 'grave implications' of the new scenario for 'development, democracy and political stability.' CBOs are locally based organizations that champion a 'bottom-up or 'people-centered' approach toward development. They are, Kumar points out, particularly vulnerable to what he somewhat surprisingly views as the 'unexpected patronage' of the donor agencies. What is most surprising is that Kamat sees this patronage as 'unexpected.' CBOs and GRDs emerged in the post-Second World War period in response to the failure of developmentalist states to ensure the basic needs of the poor—in the 1970s *the* declared development agenda of the ODAs and associated governments in the North. In this context, as well as a foreign policy concern with the spread of communism and the perceived impulse of some popular organizations and governments to take the road of social revolution toward development, USAID set up, sponsored, and financed a number (some 380 in the 1960s and 1970s) of US private voluntary organizations (PVOs) to act as private contractors of the government's foreign policy agenda. A somewhat larger number of community-based organizations in Latin America were similarly financed and sponsored.

In many cases the leaders of these CBOs were, or had been, active in women's or radical left movements, who had become disillusioned with the politics of what would later be defined as the 'old left.' These CBOs generally favored a social rather than political approach to development, with a concern for social justice and local issues. In this relatively apolitical context the CBOs were aggressively courted by both northern NGOs and ODAs such as the World Bank which, to some extent, preferred to finance and support these 'intermediary or 'local grass-roots organizations' directly rather than work through the northern NGOs. More often than not these CBOs accepted the financial support, if not tutelage, of the ODAs as a necessary evil and sometimes even as a virtue (building the capacity for self-help and social capital).

The nature of their work requires CBOs (or 'intermediary grass-roots organizations' in the World Bank's language) to interact directly with local communities on a daily basis, building relationships of cooperation and trust designed to understand local needs and tailor projects to these needs. The work of such social activists and organizations—identified by Rains Kothari as 'non-party political formations'—often was and sometimes still is looked on with suspicion by governments in the region, many of which, according to Ottoway (2003), are democratic in form but not in content ('semi-authoritarian') and the target of democratization efforts. In the interest of 'strengthening civil society' the ODAs have increasingly turned toward these CBOs rather than the NGOs as their executing agents. The dominant strategy, however, is based on partnership with local governments, CSOs and the private sector—an approach facilitated by widespread implementation of a decentralization policy (Rondinelli et al., 1989; Veltmeyer, 1997; Litvack et al., 1998).

The early history of the community development movement in the 1950s and the 1960s signified the emergence of a 'pluralist democratic culture' in many developing countries as well as a concern for local development within the framework of liberal reforms of national policy. But the dominant trend was for economic and political development based on the agency of the central government and the state. However, in the new policy environment of 'structural', free-market reform, this incipient democratic culture was cultivated by the return of civilian constitutional rule, and, at another level, by widespread policies of privatization and decentralization. With the retreat of the state from the economy and its social (and developmental) responsibilities it was left to 'civil society' to pick up the slack—in the form of emergent self-help organizations of the urban poor and a myriad of community-based and nongovernmental organizations to deal with issues of social and economic development such as health, housing, food kitchens (*comedores*, or communal dining halls), capacity-building, and self-employment. The formation of this 'civil society' was a dominant feature of the 1980s.

In the environment created by the 'new economic model' of neoliberal, free-market capitalist development, CBOs became a useful, even essential, adjunct of the policies pursued by the donor agencies such as USAID—polices designed to promote a 'capacity for self-help.' The failure of a state-led model of economic development, combined with conditions of a fiscal crisis and weakened state infrastructure, as well as a decline in state entitlements to the poor, led the donor agencies to channel an even greater share of ODA (official transfers of international resources) through CBOs and a proliferating number of NGOs. In this connection, Al Gore, on the vice-presidential campaign trail, in 1994 stated that within five years (i.e. by 1999) up to 50 per cent of USAID would be so channeled. Similarly, the *Financial Times* (July 2000) reported that the UK was increasingly inclined to fund locally-based NGOs directly, bypassing its own NGOs such as Oxfam.

The conjunction of a retreating minimalist state and the exponential increase in community-based NGOs led to the conclusion that the phenomenon was analogous to 'the franchising of the state' (Kamat, 2003: 66). In this context both the donor agencies and the IFIs recommended the privatization of both economic activity and social services—a trend that in any case was already underway—and the allocation of ODA to community-based NGOs for the same programs. Under these conditions the 'grass-roots' NGOs proliferated, as did the northern NGOs anxious to occupy the spaces left by a retreating state.

FROM THE GLOBAL TO THE LOCAL

The influx of external funds, combined with pressure to step into the spaces vacated by the state, forced many NGOs, particularly those that had 'grass-roots' or were community-based, to restructure their activities in line with a new partnership approach of the ODAs. In the process, according to Kamat (2003: 66), the organizational ethic that distinguished CBOs as 'democratic' and representative of the popular will is being slowly undermined. First of all, CBOs generally have an active membership base within the communities in which they work, be they urban slum-dwellers or poor peasant farmers. However, these 'target' or 'client' groups at the local level are themselves increasingly involved in efforts toward 'strengthening civil society'—incorporating them into decision-making processes at the local level. This form of direct or popular democracy both enthralls the donor agencies and the 'social left' but also inconveniences the former and embarrasses the latter. On the one hand, it identifies the unique strength of NGOs, which, according to the World Bank (1994), consists in 'their ability to reach poor communities and remote areas, promote local participation, and operate at a low cost, identify local needs [and] build on local resources.' On the either hand, direct democracy is inconvenient because of 'its limited replicability, self-sustainability, managerial . . . capacity, narrow context for programs and politicization' (Kamat, 2003: 66).

In this context NGOs were slowly but surely transformed from organizations set up to serve the poor into what the World Bank describes as 'operational NGOs'—private contractors of their policies that operate within 'poor constituencies' with a more or less apolitical and managerial approach (micro-project) but are not rooted in or part of these communities. First of all, the implementation of local projects calls for training in specific skills rather than a more general education that involves an analysis of social and economic policies and processes. As a result, NGO after NGO has been forced to adapt a more narrowly economic and apolitical approach to working with the poor than had often been the case. At the same time, local participation in decision-making becomes limited to small-scale projects that draw on local resources with the injection of minimal external funds for poverty alleviation—and that are not predicated on substantial social change in the distribution of, and access to, local and national resources. In this context, local community groups are left to celebrate their 'empowerment' (decision-making capacity vis-à-vis the distribution of local resources and the allocation of any poverty alleviation funds) while the powers-that-be retain their existing (and disproportionate) share of national and local resources—and the legal entitlement to their property without the pressure for radical change. In effect, the forced professionalization of the community-based NGOs, and their subsequent depoliticization, represent two sides of the same development, producing a common set of effects: to keep the existing power structure (vis-à-vis the distribution of society's resources) intact while promoting a degree (and a local form) of change and development.

EMPOWERMENT OR DEPOLITICIZATION?

According to ECLAC (1990) in its well-known programmatic statement of its alternative to the neoliberal model (*Productive Transformation with Equity*), designed, like the UNDP model of 'Sustainable Human Development' published in the same year (1990), to give the structural adjustment program a social dimension and the whole process a human face, 'participation' is the 'missing link' between the process of 'productive transformation' (technological conversion of the production apparatus) and 'equity' (expansion of the social basis of this apparatus). The World Bank had recently 'discovered' that 'participation' is a matter not only of 'equity,' as ECLAC understood it, but 'economic efficiency' (without it development projects and programs tend to fail).

As an aside, it has to be said, this recognition, stated as early as 1989, did not lead the Bank to adopt a more inclusive approach to macroeconomic policy, which, by all accounts, is profoundly exclusive, designed to benefit only those enterprises that are productive and competitive. In any case, the World Bank is in essential agreement with all of the other operational agencies of the UN system that the decentralization of government, if not the state, is an indispensable condition for both a more democratic and participatory form of economic and social (that is, *integral* or *human*) development and for establishing a regime of 'good [democratic] governance'—political order on the basis of as little government as possible but rather with what amounts to a 'system of social control' based on a consensus within civil society, at least among what the World Bank and IDB define as the 'stakeholders.' On this basis, the Bank, like the IDB, has been a major advocate of the policies of decentralization as well as the virtues of local democracy and local development (Rondinelli, McCullough and Johnson, 1989; World Bank, 1994; Blair, 1995; BID, 1996; UNDP, 1996; OECD, 1997).

The new emphasis on project implementation at the local level provided by widespread implementation of administrative (and sometimes financial) decentralization has had a number of effects. First, it has drawn attention away from the need for large-scale 'structural' change in the allocation/distribution of society's productive resources. Development projects are implemented within the spaces available within or left by the structures of economic and political power—ownership and decision-making capacity in regards to society's productive resources. Secondly, it has resulted in a programmatic focus on individual capacities, minimizing the concern for the 'structural' (social and political) causes of poverty, rejecting efforts to deal with them in a confrontational matter, promoting instead pacific ('democratic') forms of political action—consultation, dialogue, negotiations, etc.

This rather apolitical and managerial (micro-project) approach to community development draws on the liberal notion of empowerment in which the poor are encouraged to find an entrepreneurial solution to their problems. In this context, the OECD (1997: 30) defines its approach in terms of 'helping people of the world develop their skills and abilities *to solve their own problems.*' As noted above, the World

Bank adopted a strategy of 'empowerment' and 'participation'—
at least at the level of rhetoric (without any effective or specific
mechanisms for bringing about these conditions) in the interest not
only of 'equity' but 'economic efficiency.'

This entrepreneurial or neoliberal notion of empowerment is
altogether different from the critical understanding of it as a form
of alternative development promoted by CBOs. In this neoliberal
discourse on empowerment, the individual, as a repository of
human resources (knowledge, skills, capacities to decide and act),
is posited as both the problem and the solution to the problem of
poverty. Of course, this is congruent with the utilitarian notion of
the individual, when freed from government constraints imposed
by the state, as an agent of rational choice (to maximize gain and
minimize or avoid losses),[4] diverting attention away from the issue
of the state's responsibility to redistribute market-generated incomes
and the perceived need for radical change not in the direction of the
market but away from it.

The 'growth with equity' (redistributive growth/basic needs)
approach of the liberal reformers in the 1970s was focused on the role
of the state as an agency empowered to redistribute market-generated
incomes via a policy of progressive taxation, redirecting this income
to social and development programs designed to benefit not just the
poor but the whole population—to meet their basic needs. However,
at the level of the NGOs, this basic needs approach included in fact,
if not by design, a policy of conscientization—educating the poor
about structural and political issues such as the concentration of
economic and political power in the hands (and institutions) of the
elite and their own political rights. In the Latin American context
Acción Católica was particularly oriented this way on the basis of
Liberation Theology, implemented at the level of extension work in
the form of pastorals. However, from the perspective of the donors,
this approach was problematic and even politically dangerous (that
is, destabilizing) in that it could—and in different contexts did—turn
the poor to reach beyond institutional and policy reform (and 'self-
help' micro-projects) toward more radical forms of change based on
collective action, even social revolution.

The issue for the poor in this context was whether they should
be empowered as individuals to take decisions related to local 'self-
help' development (basically how and where to spend poverty
alleviation funds) or as part of a collective or community to take
direct action against the structure (and holders) of economic and

political power. There is a significant political dimension to this issue. That is, does empowerment of the poor necessarily entail a relative disempowerment of the rich—forcing them to give up some of their 'property' (a share of society's productive resources and associated incomes) and to share with the poor their decision-making capacity or power? The politics of this question was clear enough, establishing for NGOs the role that they would come to play—not the role they would take for themselves but that which they were cast into as private contractors of public policy.

In terms of actual developments from the 1970s the effect has been not to empower the poor (increase their decision-making control over conditions that directly affected their livelihoods) but rather to depoliticize grass-roots organizations of the poor—inhibiting the political mobilizing of forces of opposition to the 'system.' Poor communities have been 'empowered' to take decisions as regards how to spend the miserable and inadequate poverty alleviation funds that come their way; and this in exchange for a commitment to accept the existing institutionality and the macroeconomic policies that supports it.

Studies in different countries as well as subsequent practice confirm this practice and the role of the NGOs in regard to it. For example, Mirafab (Kamat, 2003: 69) traces the conversion of Mexican NGOs from organizations geared toward 'deep structural change through consciousness, making demands and opposing the government' into organizations aimed at an 'incremental improvement of the living conditions of the poor through community self-reliance.' This process was not unique to Mexico. Indeed, in cases too numerous to mention, community-based NGOs moved away from empowerment programs that involved the political organization of the poor based on conscientization (education about unfair government policies or inequitable social structures). Instead, at the behest of the donors, NGOs turned toward a 'skills training' approach to the mitigation of poverty by providing social and economic inputs (social capital) based on a technical assessment of the needs, capacities, and assets of the poor.

The dynamics of this conversion process vis-à-vis the role of the NGOs can be summarized as follows. 'Operational NGOs' (to use World Bank language) which established instrumental relationship with their constituencies in the marginal communities of rural and urban poor allow development experts to proceed as if the demands of the people are already known and predefined—demands such as for roads, electricity, midday meals, birth control for women, micro-finance, and poultry farming, to name but a few. In this context, Kamat (2003: 65) notes, empowerment and participation are stimulated by NGOs and their donor agencies even as their practices are increasingly removed from the meaning of these terms, which is to say, they are decapacitated or disempowered in regard to bringing about the changes needed to improve their access to society's productive resources.

The popularity of micro-credit or micro-finance projects in the practice of development can be understood in a situation wherein the state is no longer primarily responsible for creating employment, let alone improving the access of the poor to society's productive resources such as land. In the early 1980s there was a strong push to privatize the means of production and also to deregulate markets, liberating the private sector from government constraint as well as emphasizing its role in regards to economic development. In this climate even the state's responsibilities and funding in the area of social development (education, health and welfare, and social security) were cut back, shifting the former to the level of local governments and cutting back the latter in the interest of balancing the government's national accounts and budget. Empowerment of the poor, as noted by the OECD and echoed by USAID and other donor organizations, in this context is defined as and means self-help—to assist GROs in helping themselves.

Rather than assisting the poor in improving their access to society's productive resources, such as land (natural resources), financial capital (credit), or physical capital (technology), the poor are expected (with assistance, of course) to build on their own social capital—to enhance their own capacities vis-à-vis their livelihood security, achieving the sustainability of their livelihoods (UNRISD, 2000).

Micro-finance or credit programming and projects[5] are well suited to the neoliberal agenda in which risks are shifted to the individual entrepreneurs, often poor women who are forced to compete for limited resources and opportunities in a very restricted market environment. The promise of livelihood security—and local development—thus translates into optimal utilization of one's own capacities and resources rather than working against the system. In this connection, Kamat (2003: 65) concludes that the 'democratization' that NGOs represent is more symbolic than substantive. For the most part they are engaged in producing a particular kind of democracy that coincides with a neoliberal economic context. In this context,

Heloise Weber (2002: 145), research fellow at the University of Warwick's Centre for the Study of Globalisation and Regionalisation, observes that micro-credit, particularly in its Bolivian paradigmatic form, is a strategy initiated *from the outside* ('at the level of global institutions') as a means of advancing the globalization agenda—'a tool that facilitates the imperatives of globalization'—and, she adds, a tool used 'for its global governance implications.' One of these implications, as she sees it, is that

> from the perspective of the architects of global development the harmonization of local social policy at the global level . . . provides for a coherent set of tools that may facilitate as well govern the globalization agenda. The microcredit agenda (and thus, the 'poverty alleviation' strategy of the World Bank—'Sustainable Banking with the Poor') . . . is conducive to facilitating policy changes at the local level according to the logic of globalization . . . while at the same time advancing its potential to discipline locally in the global governance agenda. (Weber, 2002: 146)

As a policy initiated not at the national (or even local) level but, as Weber notes, at the level of 'global institutions' (the World Bank, etc.), micro-credit or finance is an 'explicating example' of what Deacon (2000: 213) has referred to as the 'supranationalization of local social policy.' In this sense, micro-credit, as the dominant strategy of local self-help development, based more on the social capital of the poor than an infusion of 'Social Funds' (the most popular means of implementing micro-credit programs), also has 'critical implications for political struggle' (Weber, 2002: 146). For one thing, it undermines the 'democratic provision' (Latham, 1997).

CONCLUSION: NGOS AS AN ALTERNATIVE TO SOCIAL MOVEMENTS

The push toward liberal democracy over the past two decades is part of a good governance strategy pursued under the aegis of the World Bank and other international development organizations. Other elements of this strategy include (1) democratizing the power relationship between civil society and the state; (2) strengthening 'civil society' in regard to its capacity for participation in the formulation of public policy; and (3) empowerment of the poor via the accumulation of their 'social capital'—building networks in support of a self-help strategy.

The major means for bringing about democratization in this form has been decentralization, a policy instituted by many countries in the 1980s or 1990s. Decentralization has taken diverse forms, but most generally involves a delegation of government responsibilities and policymaking capacity from the center to lower levels of government. Ironically, it was Agosto Pinochet in the early 1970s who pioneered this policy, as well as the package of 'sweeping economic reforms' used by the World Bank to construct its neoliberal program of structural adjustment reforms. In regard to this policy of decentralization Pinochet spoke of 'teaching the world a lesson in democracy'—what the World Bank (1989; 1994) has come to define as 'good governance': rule by consensus engineered via the participation of local communities in decisions relating to conditions that directly affect them. The crux of this policy is 'popular participation,' conceived of by ECLAC (1990) as 'the missing link' between the neoliberal concern with 'productive transformation' and the principle of 'equity' promoted by structuralists and reformists. In the 1980s the notion of 'popular participation' was enshrined in the notion of 'good governance' and treated as a fundamental principle in both project design and the delivery of development assistance and government services. Popular participation in this context is defined as a matter of 'equity' and 'efficiency' as well as 'good governance.'

But this concern for good governance is not what it appears to be—a matter of popular participation. Behind the notion of 'governance' is a fundamentally political concern to establish the conditions needed to implement the new economic model of free-market capitalist development—to ensure the capacity and the political will of national governments to 'stay the course' (structural adjustment, globalization) and thereby the stability of the new world economic order. Just as important is the operational and political need to subjugate local (national) economies and 'emerging markets' to the dictates of global capital. This is the agenda of US or Euro-American imperialism. It is this agenda that defines the ideology of globalization and the agency of organizations for international development. Both globalization and development as geostrategic 'meta-projects' can be unmasked as disguised forms of imperialism, raising serious questions about NGOs in the process.

It can be concluded that these organizations play a critically important role in advancing the imperialist agenda. In the 1970s many were converted into frontline 'development' agencies—to spread the gospel of the virtues of social and political reform and, within the context of local development micro-projects, help offset growing pressures for revolutionary change. In the 1980s, in a different context (external debt crisis, implementation of a new economic model, privatization and state reform, democratization), the NGOs once again were enlisted in the World Bank's declared 'war against poverty' as agents of (popular) democracy and (alternative) development—as partners in the development enterprise as well as a non-confrontational approach to social change.

This project was advanced, and the process consolidated, in the 1990s, creating conditions that facilitated the workings of Euro-American imperialism. A critical factor in this consolidation was the creation of client regimes committed to a neoliberal model of capitalist development and globalization. But another, just as critical, factor was the incorporation of 'civil society' into the development and democratization process. One part of civil society, in the organizational form of social movements, engaged the political project of opposition to neoliberalism and globalization—mobilizing the forces of resistance against neoliberalism into a popular movement or (in Gramscian language) a 'counter-hegemonic bloc' of 'popular power.' However, another part, in the form of NGOs, has been complicit with Euro-American imperialism, providing it with important ideological and political services. The actual intent of these NGOs is not the issue. In many cases the individuals involved genuinely believe that they are acting in the interest of the local communities, providing 'the poor' with essential services and benefits. Nevertheless, we need to look at whose interests are in fact served by their 'actions' as strategic partners of the international organizations. And the critical question, one that we address below in our case studies of Bolivia and Ecuador, is how NGOs play into the relationship of the state to the social movements and the struggle for political power.

2
From Popular Rebellion to
'Normal Capitalism' in Argentina

Between December 19 and 21, 2001 a massive popular rebellion overthrew incumbent President De la Rua amidst the greatest street battles and highest casualties (38 protestors were assassinated) in recent Argentine history. Major demonstrations and street blockades took place throughout the rest of the country, in an unprecedented alliance between the unemployed, underemployed workers, and a substantial sector of the middle class which had just been defrauded of its savings. In quick succession three congressional aspirants who sought to replace De la Rua were forced to resign. From December 2001 to July 2002, the burgeoning popular movements were a power in the streets and a visible presence in all provinces, blocking highways as well as the major boulevards of Buenos Aires and provincial capitals. It is estimated that up to four million participated in demonstrations out of a potentially active population of 30 million (Argentina's total population is about 38 million). Numerous writers on both sides of the spectrum spoke of a 'pre-revolutionary situation,' they wrote of 'dual power' between the '*piqueteros*,' neighborhood assemblies, and the 'occupied factories' on the one hand, and the existing state apparatus on the other. There is no question that the principal arms of the state apparatus (the judiciary, the police, and the armed forces) as well as the traditional parties, politicians, and Congress, lost their legitimacy in the eye of a majority of Argentines in the events leading up to and immediately after the uprising of December 2001.

The most popular slogan '*Que se vayan todos!*' ('Out with the politicians!') reflected the general hostility of the public toward the major parties and political institutions. Yet 17 months later, over 65 per cent of the electorate voted and the top two candidates were from the Justice Party (Peronist), including Carlos Menem, president between 1989 and 2000, the main culprit for the collapse of the economy and the impoverishment of millions of Argentinians. Facing a resounding defeat in the second round, Menem withdrew and Nestor Kirchner, with a little over 21 per cent of the vote, became president.

Barely two and a half years after the uprising President Kirchner was enjoying a 75 per cent approval rating, the support of the three major trade union confederations, the backing of the human rights organizations (including the militant *Madres de la Plaza de Mayo*), vast sections of the middle class, and many important '*piquetero*' organizations of the unemployed, in addition to the backing of the IMF (with some opposition).

The abrupt and profound political transformation raises a series of important theoretical and practical questions about the nature of the popular movements and uprisings, both their achievements and limitations. More specifically the Argentine 'political transformation' raises several other questions concerning the legitimization of political institutions which were clearly discredited; the political strategy of a neoliberal regime in a time of widespread rejection of failed neoliberal policies; the reformation' of the Peronist Justice Party—subsequent to the December 2001 uprising; the multi-polar international economic strategy (ALCA, Mercosur, the Free Trade Pact with the EU; bilateral relations with Venezuela, Brazil, and China) designed to secure export markets.

If significant political *changes* have taken place, they occur in a context of substantial continuities on socioeconomic structures and policies, which have had only slight impacts on the class structure, including unemployment, incomes, and poverty. Moreover, the most serious obstacles to any sustainable, equitable, and vigorous economic development remain untouched: foreign debt payments, capital flight and overseas deposits, privatizations, and disinvestment by foreign and domestic owners of strategic enterprises. The big question facing the 're-bourgeoisification' of Argentine politics is embedded in this heterodox regime—how deep and how far can the Kirchner administration proceed with political and social changes in the face of Kirchner's commitment to strategies and institutions linked to the past? Specifically, is Kirchner's vision of 'normal capitalism' growing out of the reigning neoliberal politico-economic configuration viable? Leaving aside the vagueness and ambiguity of the term (what is 'normal'?), is there any way that Kirchner can cultivate a 'national bourgeoisie' in the face of neoliberal policies, free trade agreements, a limited domestic market, and energy and infrastructure bottlenecks resulting from the previous privatization policies?

To address these questions we will examine the changes and continuities during the Kirchner regime. Our aim is to analyze the underlying logic and dynamics of the regime and to understand its future course, potentialities, and constraints.

POLITICAL TRANSFORMATIONS

Kirchner has carried out an important series of changes in the judicial, military, and law enforcement institutions. That is to say, he has successfully replaced the corrupt 'automatic majority' of the Supreme Court justices appointed by President Menem with a group of respected jurists. He has forced into retirement many of the top generals and police chiefs with dubious human rights credentials, many involved in illicit contraband, kidnapping, and extortion activities. He has led a successful fight to repeal the amnesty granted by Presidents Alfonsín and Menem to the generals involved in the mass murder of 30,000 Argentinians during the 'dirty war' years (1976–82). He has been active in demanding that Congress clean up its high level of bribe-taking (especially the bribes taken in 2001 to pass anti-labor legislation). Through these policies, Kirchner has partially relegitimized public institutions by giving them at least a semblance of honesty, accountability, and responsiveness to human rights concerns. Equally important, through style and substance he has legitimized the presidency as a valid interlocutor with sectors of the popular movements, human rights groups, trade unions, and the international financial institutions. As a result Kirchner, as of March 2004, had a high popularity rating going into his second year in office. Nevertheless, these important political changes did not significantly affect the nature of the public institutions or their political-class allegiances.

The 'reformed' military is a case in point. While some of the high-ranking officials were retired, most of those who took their place were of the same school of authoritarian politics—and profoundly hostile to the re-trial of the genocidal generals. This became evident during Kirchner's visit to ESMA, the former Naval Academy converted into a museum recognizing the victims of the military terror. The same is true in the judiciary and police: changes in personnel at the top have not changed the 'rules' and context of operation of the officials. Many of the judges are still part of the old order and the police still engage in corrupt and violent activities. The partial 'house cleaning' is not a continuous process and the conditions exist for the reproduction of the old system, once circumstances and mass pressure recede.

The repeal of the amnesty has not led to a rapid re-trial of the genocidal military officials. Human rights observers claim that there will be a prolonged process of litigation prior to the trials, in part because a substantial sector of the pro-Menem Peronists are important state governors and Congress members and continue to oppose human rights trials. While Kirchner has made significant changes at the top, he has not changed the structural linkages between the political institutions, his political party (the Peronists), and the neoliberal economic elites—both foreign and domestic—who continue to control the economy.

Some policy changes have taken place under Kirchner, weakening the long-standing authoritarian elites. The president intervened in Santiago de Estero to remove the Mafioso governor whose family had ruled with an iron fist for over half a century. In line with his effort to build his own independent political base—against the rightwing Peronists—Kirchner has promoted what he calls 'transversal alliances,' coalitions that cut across existing party and social movements. While still heavily dependent on Peronist governors, the Kirchner project envisions a new political party of the center-left based on a return to the national-popular politics of the earlier Peronist era, but with less corruption and repression.

During its first year in office the Kirchner regime was generally tolerant of *piquetero* activity—including street blockades—thus avoiding violent confrontations that might alienate his supporters among the human rights groups and re-ignite mass protests. While the Kirchner regime was initially respectful of the democratic rights of the *piqueteros* to protest, he refused to rescind the political trials of 4,000 activists arrested during the previous regime. Moreover, his provincial political allies continued to repress mass protests savagely, jailing and injuring many in San Luis, Santiago de Estero, Salta, and Jujuy. Worse still, Kirchner's Minister of the Interior promises to enforce a court ruling of April 2004 criminalizing *piquetero* street blockades, a measure that united the entire *piquetero* movement in a massive protest in May 2004. Kirchner's first year in office was based on a realistic assessment of his precarious position, the proximity of the December 2001 uprising, and the need to convert his general popularity into an organized and 'organic' base of support. Kirchner has moved with great astuteness in this regard, balancing his liberal economic policies with human rights measures and widespread but minimalist welfare policies. He has complied with most of the IMF commitments but has rejected increases in budget surpluses and higher payments to private bondholders, thus keeping the international financial institutions at bay while creating the popular image of being independent of the IMF. He has proceeded to satisfy symbolically many of the demands of the human rights groups over past violations, without threatening the current military officials. He has not reversed the privatizations, but temporarily froze the prices for energy, electricity, and other public services, a policy he subsequently reversed. Kirchner retains most of the subsistence 'labor contracts,' but has not increased payments beyond the $50 a month rate and has eliminated over 20,000 recipients who the regime claimed were ineligible, despite protests to the contrary by some *piquetero* organizations.

Kirchner's most ambitious and successful social program is in the area of pharmaceuticals: the government guaranteed public provision of drugs at a 90 per cent discount in the primary care clinics for low-income families, covering 15 million people. In addition, the government claims to provide anti-viral medications for AIDS victims. The Minister of Health, Gines Gonzalez Garcia, claims that the new Generic Prescription Law covers 82 per cent of the drugs prescribed in Argentina, enabling four million Argentines to access drugs previously beyond their purchasing power.

Given the high level of abject corruption in the state apparatus, Kirchner was obliged to make changes in personnel in order to have a political instrument to pursue the 'normalization' of capitalist development. For example, given the connivance of high-ranking police authorities with gangsters and criminal police involved in widespread extortion, blackmail, and kidnappings of business people, it was impossible to secure investors and new investments. While the Supreme Court was controlled by corrupt judges at the beck and call of ex-President Menem, there was the constant threat of any change being ruled 'unconstitutional.' The key to Kirchner's political, judicial, and military reform is found in his desire to make the state over in the image of 'normal capitalism' without unmaking its capitalist and even neoliberal character. Unquestionably, some of the changes were positive—as far as they went. But it is also the case that as Argentine capitalism once again goes into crisis even these 'reformed' state institutions can serve to repress and block necessary changes.

ECONOMIC PERFORMANCE AND SOCIAL CONDITIONS

The recovery underway in Argentina reflects the profound class bias in the Kirchner regime. Foreign trade based on agro- and mineral exports has boomed, but has led to few if any 'trickle-down effects.' International trade negotiations over ALCA have taken place and some disagreements have been aired, but mainly affecting Argentine agro-exporters and not with labor, environmental, or social interests.

Between March 2002 and January 2004 industrial production grew by 33 per cent. Economic growth in 2003 was 8.7 per cent. According to official statistics, unemployment declined by 5.9 per cent, from 21.5 per cent to 16.3 per cent. Early projections based on January to March 2004 suggest that GNP would continue to grow at a similar rate for most of that year unless energy shortages curtailed growth. In large part this growth is based on favorable internal and external circumstances rather than on any structural changes. The prices for most of Argentina's principal agro-mineral exports were at record highs in 2003 to mid-March 2004—oil, meat, grains, soybean at near-record highs, providing a trade surplus of over 5 per cent of GNP. This allowed the regime to meet the 3 per cent surplus with the IMF and facilitated financing the partial economic recovery and the work plans (unlike the case in Brazil where the Da Silva regime allocated a 4.25 per cent budget surplus to pay off creditors at the expense of the local economy and the unemployed).

In large part the progressive economic image of the Kirchner regime is an artifact of the *context* in which it acts more than in the substance of its policies. Given the ultra-liberal policies currently practiced in Brazil, Mexico, Uruguay, Peru, Ecuador, Chile, and elsewhere, and given the devastatingly destructive give-away policies of his predecessors, it is understandable that many journalists, leftist intellectuals, human rights activists, and others perceive Kirchner's vision of 'normal capitalism' as progressive. If one adds the attacks from the far right in Argentina (sectors of the military, the *Nación*, financial speculators, and provincial governors linked to ex-President Menem) and sectors of the IFIs externally it is plausible to argue that, against them, Kirchner is 'progressive,' but *only* in a very limited sense, in both time and space.

The growth sector in exports across the board is in large part because of the devaluation, the high demand for Argentina's commodities, and the low starting point for measuring the recent recovery. Between 2001 (the year of the financial economic collapse) and 2003, Argentine exports grew between 11 and 124 per cent depending on the sector—the biggest growth taking place in agriculture and petroleum commodities, which did not require new investments. The devaluation also stimulated the growth of local industries, as imported capital and finished goods became too expensive. The net result was significant trade surpluses. The same pattern of trade surpluses is evident in 2004: for the first two months of 2004, Argentina realized a $1.7 billion trade surplus. In large part the Argentine export boom acts as a motor to economic recovery because its domestic market is still very depressed.

Export growth has bypassed the majority of Argentinians for several reasons. The export growth sectors are very capital-intensive and employ a very small number of workers (mechanized agriculture and petroleum). The income and profits from the export sector accrue to a very small class of foreign capitalists and local agro-oligarchs who transfer a substantial percentage of their earnings abroad, lessening the so-called 'multiplier effect' on the rest of the economy. Overall inflation figures of approximately 4 per cent were highly deceptive as they reflected prices of durable and capital goods; prices of meat and other basic consumer food items rose by nearly 20 per cent, thus prejudicing workers, public employees, and unemployed workers whose wages were frozen.

For Argentinian speculators 2003 was a glorious year as stocks soared by over 100 per cent. Backed by the IFIs, foreign bondholders still refused to accept a 75 per cent reduction on their defaulted paper from previous regimes. The financial system showed some recovery as deposits rose 50 per cent since the middle of 2002, yet more than 90 per cent are held in short-term accounts of less than 30 days. The financial and industrial sectors of capital show little inclination to make long-term, large-scale investments, which would sustain growth. Instead, they continue to send their earnings abroad and in some cases disinvest. During 2003 net capital inflows were a negative $3.8 billion. Almost all 'growth' in 2003 was based on activating existing unused capacity, which still remains a serious bottleneck in the economy. A particular high-risk obstacle to growth is the foreign monopolies in the country's electricity, gas, water, and telecommunications industries, which have not made major investments as agreed to during the signing of the privatization contracts. As a result, even the one-year growth of 2003 is at risk as the monopolies have created gas and electrical 'shortages' to increase the rate of profits via increased prices, thus forcing reductions in producer and consumer usage.

Kirchner's continuation of the fundamental structures of economic power is creating serious economic obstacles to even medium-term growth. Worse, the social crisis has deepened or practically stayed the same as in the past.

CONTINUITY OF THE SOCIAL CRISIS, 2003

While GNP increased in 2003 it was still 11 per cent below the level of 1997. While per capita GNP increased by 6 per cent, it is still 17 per cent below what it was seven years previously. While unemployment declined 5 per cent it was still a hefty 16.3 per cent. Moreover, the number of part-time workers increased from 13.8 per cent in 2002 to 16.6 per cent in 2003. If we combine part-time and unemployed workers, the net gain between 2002 and 2003 (the 'recovery' year) the change for the better is very marginal—2 per cent. The 'recovery' has not had any major impact on the working classes in relation to the problem of employment. If we subtract the two million beneficiaries of the $50 a month work plans it is evident that the worst of the crisis still affects the bulk of the working class. Data related to salaries, levels of poverty, and indigency also substantiate the argument that Kirchner's continuities in economic policies have not had any positive impact on the worker and salaried classes.

Between 1997 and 2003 average salaries declined 22.4 per cent. There was a steady but gradual decline until 2001, which was followed by a sharp fall in 2002 of 17 per cent. Salary deterioration continued through 2003, falling 0.8 per cent during the Kirchner presidency because economic growth failed to have any spill-over effects on public sector workers whose salaries were frozen (until May 2004) and whose trade union leaders supported Kirchner's policy of holding back wage demands to spur growth. The fall in purchasing power during the 2003 'economic recovery' was even greater if we look at the above-average increase in the price of basic foodstuffs, which form a larger proportion of the family basket for working-class families. Clearly the export-driven recovery, compliance with the IMF's stabilization program, and the high returns to speculators in the Argentine stock exchange came at a continuing high price for the majority of wage and salaried Argentines.

The most striking illustrations of Kirchner's elite-based recovery is found in the year-to-year data on poverty levels and indigency.

By the end of Kirchner's first year in office over half the population still lived below the poverty line (51.7 per cent). Despite high growth, poverty levels declined by only 2.6 per cent between 2002 and 2003, frustrating any reading of Kirchner's first year as an 'economic success story.' Worse still, the level of indigency continued to be extraordinarily high—25.2 per cent in 2003, actually rising 0.5 per cent over 2002.

THE NATIONAL BOURGEOISIE AND THE ENERGY MNCS

The explanation for the inverse relation between GDP growth and salary, poverty and indigency deterioration is found in the behavior of the state and the national bourgeoisie: the state for allocating billions in debt payments to the international lending agencies instead of job-creating public investments; for continuing to allow lucrative privatized enterprises like Repsol Petroleum to transfer billions to their home office; for not taxing the $150 billion in overseas holdings of the Argentine elites; for failing to channel foreign exchange earnings of the agro-mineral elites into job-creating domestic production. In short, Kirchner's commitment to the neoliberal, imperial-centered model of his predecessors does not provide the political-economic instruments, resources, and capabilities to tackle deep structural configurations directly, which generate poverty, indigency, and declining living standards.

The 'national bourgeoisie' taking advantage of Argentina's import constraints has launched an economic recovery of sorts. Taking advantage of the large reserve of unemployed, Argentine capital has expanded production by exploiting low-paid labor and longer working hours to increase output, without any significant new investments or technology. In effect, the industrial recovery has taken place via the activation of unused existing capacity. There is little sign of any large-scale, long-term new industrial investment. Once this 'easy development' runs its course, Kirchner's 'national bourgeoisie' shows few signs of sustaining development. In the meantime, the strategic, foreign-owned economic sectors are moving hard and fast to capture whatever economic surplus emerges from the recovery. Foreign creditors and the IMF are demanding an increase in the valuation of the deflated bonds. The privatized gas and electrical companies, which reaped exorbitant profits between 1996 and 2001, failed to comply with their investment agreement with the government. As a result, with the recovery, the country's industries and consumers face a severe energy crisis. In April 2004 the state agency in charge of regulating the electricity wholesale market was forced to lower the voltage to prevent greater blackouts induced by the big electrical companies (*Financial Times*, April 5, 2004 p. 4). During the last week of March 2004 more than 30 industries were affected by blackouts. According to the *Financial Times*, the energy crises could reduce economic growth by 2 per cent in 2004 (ibid.). The Kirchner regime capitulated, allowing gas companies to raise prices by 100 per cent over the next 15 months. While the presidential claims that smaller companies and residential consumers would not be affected, most likely the larger industrial users will pass on to consumers the higher energy costs. Even more important the regime's energy price increase accepts the 'principle' of raising rates and rewarding corporate blackmail. Kirchner's regime has privileged the position of the privatized foreign-owned petroleum companies, thus rejecting popular demands to renegotiate the unfavorable terms of the privatization.

The petroleum and gas industries are one of Argentina's principal export and foreign exchange earners. Beginning with his earlier period as governor of Santa Cruz province, Kirchner has been a staunch supporter and promoter of the privatization of petroleum and a close ally of the Spanish multinational owners, the Repsol Petroleum Company. The long-term deep structural links between the Kirchner regime and the foreign-owned petroleum corporations represent a triple threat to Kirchner's vision of 'normal capitalism.' In the first instance, the demands of the energy sector for higher charges to consumers (both industrial and household) will certainly be inflationary and unpopular and will reduce Argentine producers' 'competitiveness' on international markets by raising the cost of production. Moreover, by raising rates Kirchner's image as an 'independent national statesman' is called into question. Kirchner's forceful but inconsequential accusations that the foreign-owned energy companies failed to invest since 1996 are true but is not followed by any positive actions or even investigations, let alone renegotiations of the privatization contracts. Secondly, the energy 'crisis' undermines Kirchner's vision of 'normal capitalism' based on an alliance between agricultural and energy exporters and the national industrial bourgeoisie. The energy shortage has already led to cutbacks in industrial production as the foreign MNCs flex their muscles.

Most energy experts familiar with the tactics energy corporations employ to raise rates question the whole idea of a 'crisis,' particularly when the energy companies reduce production 'for maintenance reasons.' This tactic was used by the MNCs in California, New Zealand, Australia, and many other regions, particularly in periods of economic recovery and growth when demand is high and the state is unwilling to intervene against the energy enterprises.

Finally, the energy crisis has generated conflict with Argentina's neighbors, namely, Chile, Uruguay, and Bolivia. In response to the

MNC-induced 'energy crisis,' Kirchner has reduced the export of gas to Chile and Uruguay in order to supply local industries and consumers. In addition, Kirchner signed an agreement with Bolivia's President Mesa for additional shipments of gas, so provoking the wrath of the Bolivian people who demanded a new hydrocarbon law favorable to the Bolivian state before any new agreements are reached. The short- and middle-range impact of the energy crisis has brought to the fore a major *contradiction* between Kirchner's regime and the majority of his supporters: the conflict between the privatized foreign-owned MNCs, which Kirchner supports, and the demands of the people for an investigation, renegotiation, and re-nationalization of strategic industries because of their high profits, abusive practices, price gouging, and minimum impact on employment, poverty, tax revenues, and national industrial growth. The energy issue, in all its ramifications in terms of both its direct costs to growth and living standards and its symbolic meaning as a reminder of the continuation of the previous era of corruption and pillage, is central to Argentina's future development and the stability of the Kirchner regime.

Kirchner's energy agreements with Venezuela, Brazil, and Bolivia may temporarily relieve the gas supply problems, but those agreements do not deal with the related problem of electricity generation. With demand growing at 1,000 megawatts a year and with spare capacity at only 3,000 MW, with no new plants scheduled by the foreign-owned electrical companies, Kirchner faces a severe political and economic crisis, particularly with the economic recovery and the growth of industrial production. Kirchner is a 'victim' of his own ideology (neoliberalism), his structural relations (with the foreign owners of energy and electrical plants), and political alliances (with the Peronist Party—the authors and executioners of ex-President Menem's corrupt privatizations and subsequent pillage).

As Kirchner began to capitulate to some of the pressures for price increases from the foreign-owned public utilities his popularity started to decline from 80 per cent at the end at 2003 to 60 per cent in June 2004. Kirchner knows full well the rules of neoliberal capitalism. The only way to secure new investments from the foreign owners of strategic industries is to increase their rate of profits—above that of 'normal capitalism.' Without sacrificing the living standards of workers and increasing costs for local producers in order to increase profits demanded by the foreign MNCs, Kirchner will not get a promise of new investments from the owners of electricity, gas, water, telecommunications, and infrastructure. Higher costs resulting from Kirchner's capitulation to high profit margins demanded by the strategic sectors will not only lower his popularity, but also his legitimacy as a 'national' leader. Moreover, it will lower growth, especially in labor-intensive manufacturing sectors—and hence fail to reduce unemployment and poverty.

Argentine 'national' capital invests by borrowing—not from reinvesting its profits in productive sectors (income is sent abroad or invested in high-return, local short-term bonds). The current growth is not based on new investments but in activating unused capacity. When installed capacity equals demand, growth will stall unless there are new investments, which the Argentine bourgeoisie will realize only if they have access to credit. Both the national and foreign financial sectors are not offering credit unless the Kirchner regime 'compensates' them for their speculative losses incurred subsequent to the devaluation. Overseas private bankers are refusing financing until Kirchner capitulates to their demands to raise the payments offered to the private overseas bond speculators.

In the short term Argentina will continue to 'recover', based in part on the extraordinary boom in agro-exports, high petroleum prices, and the reactivation of industry from its cataclysmic decline between 1998 and 2002. But the underlying structural and ideological foundations, which produced the crisis and popular uprising, are still in place. Moreover, the tendency is for the government to move toward a greater accommodation with the elite foreign beneficiaries of the neoliberal model. First and foremost, the regime has *legitimized* the privatizations and called for greater foreign investments in the 'development' of basic infrastructure and exploitation of strategic resources. In keeping with this priority the Kirchner regime is gradually (and with occasional populist demagogy) moving toward implementing economic policies that increase profitability for the firms, even at the cost of living standards. The banking reform and debt agreement with overseas speculators will weaken the capacity of the regime to meet the social demands of the majority of Argentines living below the poverty line. The unfolding reality is a series of incremental concessions to foreign owners of strategic economic sectors and overseas speculators designed to facilitate financing of local investments. Kirchner's vision of the decent face of 'normal capitalism' appears to be a clay mask, which as it falls away reveals the smirking face of the old 'pillage and run' capitalism of the too recent past.

THE SOCIAL MOVEMENT:
PIQUETEROS, NEIGHBORHOOD ASSEMBLIES, AND *PUNTEROS*

Most writers cite the December 19–20, 2001 uprising ousting President De la Rua as a 'turning point' in Argentine history. There is no doubt that the mass, largely spontaneous mobilization led to spectacular challenges to the existing political order, at least in the short run. Throughout the country, neighborhood assemblies met in previously quiescent lower-middle-class and even middle-class *barrios* demanding the return of their savings. For the first time there were joint marches of the organized unemployed and sectors of the middle-class neighborhood associations. The clientele patronage system through which the Peronist bosses controlled poor neighborhoods was broken, as new autonomous unemployed movements emerged, took to the streets, blocked traffic, and negotiated concessions directly from the state. Public debates over public polity involving over one-fifth of the entire adult population temporarily replaced the elite wheeling and dealing in Congress, which passed for 'democratic politics.' The whole political class, their parties, and public institutions fell temporarily into total discredit. The populace in Buenos Aires, at one point, even stormed the Congress. Likewise in the provinces, they invaded the legislative assemblies, tossing furniture out the windows in their rage at the venality and unresponsiveness of the legislature and the party bosses who controlled the electoral processes and elected representatives. In the early weeks and months following December 2001 it seemed that a new political order, a new political discourse, a new way of 'doing politics' was emerging.

The movements of the unemployed, with their *piquetero* activists, were organizing a vast network of the poorest of the poor. The unemployed former trade union metal-workers were applying their old organizing skills learned in the factories to organize the unemployed in their new settings in the *barrios*. Road blockages had the same effect as factory-based strikes—paralyzing the circulation of commodities. Unlike the bureaucratized trade unions, the unemployed movements took their decisions in mass popular assemblies. Direct and autonomous organization free from party control seemed the order of the day. There was a 'feel' of a potential new and more responsive and fully democratic order in the making.

But it was not to be. In the course of only two years, the process of 'democracy from below' began to ebb and then went into full retreat during the first year of President Kirchner's regime (May 2003—April 2004).

What happened to the promise of a new political order 'from below'? What went wrong? What is left from the December 19–20 uprising two and a half years later?

The original strength of the popular uprising—its spontaneous, mass, autonomous character—became its strategic weakness, the absence of a national leadership capable of unifying the diverse forces behind a coherent program aimed at taking state power. Instead, the potential of the unemployed workers' 'movement' fragmented into a series of smaller movements, each led and controlled by local leaders or by small leftwing parties. The middle-class assemblies initially drew hundreds of neighbors to all-inclusive marathon discussions, which eventually exhausted their participants without leading to any formal organization, specific program or even city-wide coordination. At the same time, the small leftwing parties brought their sectarian conflicts into the assemblies, driving out many by their jargon, their maximalist programs, and their inability to solve pressing immediate problems, such as recovering the savings of middle-class depositors in frozen and devalued accounts in the foreign-owned banks.

Between December 2001 and July 2002, the mass movements, despite their division, were still on the offensive, challenging interim President Duhalde. They dominated the streets, calling into question the legitimacy of the political system. In this half-year interim, as the economic crisis deepened and unemployment soared to over a quarter of the labor force and the middle class lost over 60 per cent of their purchasing power, the three trade union confederations (from both the right and left) failed to respond to the political crisis. The unemployed workers' leaders made no effort to create a new alternative labor-based trade union. The original virtuous notion of autonomy from the old political parties became a slogan to justify the rise of local personalist leaders in each *barrio*, who undermined any effort to unify forces into a national or even citywide social movement. 'Autonomy' among a couple of student-influenced *barrios* became an excuse to turn away from politics to self-help projects. The crisis continued.

The Duhalde regime 'tested' its strength and capacity to reverse the movements via repression. In June 2002, a police inspector murdered two unarmed *piquetero* activists in a video-recorded confrontation, provoking a massive demonstration. The regime turned toward handing out hundreds of thousands of six-month work plans to *piquetero* organizations and local Peronist bosses. The *piquetero* leaders originally saw the demand for work plans as short-term solutions in response to malnutrition and growing indigency. Originally, the work plans mobilized hundreds of thousands because they presented a concrete gain in a very specific and urgent circumstance. Although other, more structural demands were included, for example the repudiation of the foreign debt and renationalization of privatized banks and strategic energy industries, over time the focus of the mass movement became the 'work plans.' 'Success,' as it was, was measured by which group or leader was most successful in negotiating the greatest number of plans in the shortest time with the least amount of paperwork. The 'work plans,' paying only $50 a month (150 devalued pesos), was far below the poverty line and close to the level defining indigency. The Duhalde regime remained discredited, under siege, but through the work plans it began to reconstruct a local apparatus to weaken the grass-roots organizations.

In the meantime, the government began a process of unfreezing and repaying at least in part the middle-class depositors, declaring itself in default on the private debt and 'stabilizing' the precarious economy. Capitalizing on splits and subdivisions among the *piquetero* movement and on the conversion of local leaders into distributors of labor plans, the Duhalde regime consolidated power at the national level, drove a wedge between the middle-class bank deposit protestors and the unemployed, and confined the influence of the popular movements to their immediate locale. In the course of his year in power (2002) Duhalde exhausted all his political credibility as the socioeconomic crisis, with over 53 per cent living below the poverty line, continued. In a shrewd move to seek to re-legitimate the political system, he called for presidential and congressional elections for May 2003. Deep political divisions within the Peronist Party led to several 'Peronist' candidates standing in the election. The extreme right wing put up former President Menem, author of the collapse of the economy, master of massive corruption, and the hate-object of the great majority of Argentinians. His main opponent was Kirchner, a former governor from Santa Cruz province, who presented himself as the reform candidate, committed to human rights, to purging the Supreme Court, the police, and other public institutions of corruption, and reviving the national economy—the return to normal capitalism—even if that meant confronting the IMF.

The Kirchner Election: Left Defeat—I

The first major political defeat of the left was in the May 2003 presidential elections. The left was as usual divided into small groups of Marxist parties which ran their perennial presidential candidates while the majority of the left called for a 'militant' voter abstention. Both factions suffered serious defeats. The electoralists got barely 1 per cent of the vote while the abstentionists failed. The voter turnout was over 70 per cent, the biggest in recent decades. Menem, who won a plurality in the first round, withdrew in the run-off as all the polls indicated a massive and humiliating defeat by close to three-quarters of the electorate. Kirchner became president.

The left, the *piquetero* leaders, and the militant human rights groups totally misjudged the changing times. 'They acted,' a coalminer leader told us, 'as if they had a bucket over their heads. Hearing their own slogans reverberating, they thought it was the voice of the people.'

After five years of recession leading to an economic depression and financial collapse, after 18 months of demonstrations and mobilizations in which the deeply divided and squabbling left was not able to change the political regime via extra-parliamentary action or even to present a unified electoral program and candidate, the mass of voters, including the great mass of slum-dwellers and impoverished middle classes, turned out to vote and place their hopes in Kirchner with his 'anti-establishment' image.

The Kirchner Presidency: The Social Movements and the New Political Project

Kirchner, taking office in May 2003, continued the work plans initiated by his predecessor Eduardo Duhalde: two million heads of household with children receive $50 a month. This continues to constitute the major social program of the government. This payment covers only a third of the cost of the basic food basket of $140 per month. Moreover the payment covers only 40 per cent of the unemployed or underemployed. The key purpose of the work plans, from their origins to the present, was never to solve the problem of malnutrition or unemployment, but to 'contain' discontent. In the beginning, the work plan program stimulated the organization of the unemployed and demands for more work plans of greater duration than the original six-month contracts.

The 'work plans' have failed to connect with the creation of new full-time jobs and thus have 'consolidated' a permanent indigent class with no future. In the great majority of cases the 'work plans' have come under the control of authoritarian provincial governors, mayors, and *barrio* bosses, who frequently hand over the payments to local clients who are neither unemployed nor in need, to the detriment of indigent families.

The unemployed workers' movements (MTDs) are divided into three: those supporting Kirchner; those giving him 'critical support'; and those opposed to him. The pro-Kirchner sectors of the MTDs (both variants) are accompanied by the three major trade union confederations (the CTA, the CGT, and the transport workers), sectors of the worker-run factories (*fabricas recuperadas*), and the major human rights groups (including the two *Madres de la Plaza de Mayo*, and the Grandmothers' Movement).

The anti-Kirchner MTDs are in turn divided into various coalitions which have frequently acted independently of each other. Sectors of the worker-occupied factories, led by the Zanon ceramic factory, are in opposition but remain largely isolated, as many of their former allies have turned to class collaborationist politics.

The deep division among the movements was visible at the anniversary mobilization commemorating the December 20–21 uprising. In December 2003, three different 'coalitions' met at rotating hours in front of the presidential palace: one coalition from 12 noon to 2 pm assembled 10,000, a second from 3 pm to 5 pm had 15,000, and the third group from 6 pm to 8 pm had 25,000. Fifty thousand unemployed demonstrators would have been a formidable expression of force, if it were not dissipated into three separate protests demonstrating the weakness of a deeply divided movement incapable of even celebrating a 'turning point' in history.

The disunity of December 2003 could be considered another 'turning point,' expressing the decline of a divided, partially co-opted movement. The 'normal capitalist' regime appeared to have temporarily consolidated its rule, once again attracting middle-class support against the militant MTDs. The militant mobilizations were focused on 'work plans' and 'decent jobs,' which have little resonance on the once rebellious but now conformist middle classes who have recovered some of their savings. The marches through the center of Buenos Aires do not draw any participants or applause. The former bangers of pots and pans are silent, scurrying on their way to their daily chores. The street protests are more likely to draw the ire of middle-class commuters and transport workers than their sympathy. A few of the leaders of the MTDs mobilize protests while preparing to run for elected office within the traditional parties.

Some of the leaders of MTDs are not unemployed workers, but leaders of political parties who lead and divide the movement. Some leaders are trade union activists with aspirations to head their unions. The tiny band of self-styled 'autonomists,' intellectual followers of Tony Negri, have virtually disappeared from the social scene, their followers joining one or another of the influential groups distributing work plans.

The movement of self-managed factories has been contained. Virtually no new factory occupations have occurred. The existing factories have come under government tutelage, and in some cases their former Trotskyist labor lawyers no longer have influence. The most striking reversal is the virtual disappearance of the middle-class neighborhood assemblies. In some *barrios*, committees still meet to discuss neighborhood problems, but the large, open-air assemblies are a distant memory. More ominously, important sectors of the middle classes have turned to rightwing authoritarian and repressive ideologies to deal with the rising crime rates. The biggest demonstrations in Buenos Aires over the past two years was a 150,000 protest—almost exclusively middle-class—in front of the Congress against crime with all the emphasis on greater repression and no mention of the immense poverty and unemployment that is correlated with crime.

During the first year of Kirchner's regime the MTDs seemed isolated, and some were controlled by leftwing parties and traditional party bosses. As the struggle against the regime ebbed, the internal struggles within the movements intensified. While previously divisions were based on competition between *barrios*, the new splits include divisions within *barrios*. For example one 'coalition' of MTDs includes the Polo Obrero, *Coordinadora Unidad de Barrios* (CUBA), MTR (one sector), and the *Movimiento Territorial de Liberacion* (MTL). The CCC, (Combative Class Current) is another unemployed group linked to the Revolutionary Communist Party. The MTR (Martino section) is another division from the Teresa Rodriguez Movement. Even the human rights groups have drawn divisionary lines. The *Madres de Plaza de Mayo* no longer allow their university to become a meeting place for the anti-Kirchner *piqueteros*, while supporting others which do support him such as the *Barrios de Pie, Patria Libre*, and sectors of the MTD, *Anibal Veron*.

The net effect of the divisions was to weaken the attractiveness and organizational capacity of all *piquetero* organizations. The May 1, 2004 (May Day) meeting was a case in point and the result was disunity and a low turnout, as each *piquetero* grouping 'celebrated' in isolation from the rest and attendance was only a fraction of the previous two years. Apart from personal rivalries, longstanding conflicts and tactical differences, the main division running through all the unemployed movements is political—specifically, their position and response to the Kirchner regime. Interviews with a wide range of *piquetero* militants and leaders in the greater Buenos Aires area in April 2004 reveals a deep crisis in the movements, a sharp division in outlooks on the regime, the level of militancy among unemployed workers and future political perspectives.

Two cases suffice to illustrate these political divisions: the Unemployed Workers Movement—*Anibal Veron* (MTD-AV); and the Teresa Rodriguez Movement (MTD-Martino). It should be mentioned that even within these movements there are internal differences and possible future divisions.

The MTD-AV, like many of the *piquetero* organizations, begins its analysis by providing an overview of the 'new situation.' They argue that the Kirchner regime is a hybrid of the national bourgeoisie and multinational corporations, which is substantially different from the previous 'neoliberal' regimes. They cite the regime's progressive policies on human rights, the changes in the military (the retirement of 40 generals), standing up to the IMF, support for Castro and Chavez, a review of retrograde labor legislation, and its less repressive policies toward public protests. To these 'positive' assessments of Kirchner, the MTD-AV adds a 'negative appraisal' of the current state of mass movements. They speak of the internal splits in their own movements, the declining numbers turning out for protests, the general 'disorientation' in the masses faced with the new regime's policies, and the 'disarticulation' of the movements as a result of regime-sponsored work plans resulting in the loss of a consensus among unemployed movements, making it difficult to struggle. As a result of these factors, the MTD-AV has adopted a position of critical support for the regime, moved away from conflict and confrontation toward discussion and negotiations, mainly over the number of work plans and local project funding. This shift in policy has led to the virtual disintegration of the small 'autonomous-decentralist tendency' in the movement organized by university disciples of Professor Jon Holloway of Edinburgh University. The issue of *political power*, more specifically, state power, and the potential economic and political benefits are key factors in shaping political attitudes among the unemployed. The leaders of MTD-AV argue that there are three possible approaches to the Kirchner regime: the position of 'Workers Pole' (controlled by the Workers Party—Partido Obrero (PO) which claims 'nothing has changed' and continues with the politics of confrontation; the politics of conciliation and subordination pushed by D'Elia of Matanzas; and Barrios de Pie and their positions of 'political independence while avoiding confrontation and relying on negotiations'—in effect, critical support. The MTD-AV argues that, given Kirchner's 60 per cent public support, 'there is no basis for confrontation.' The MTD-AV argues for pressuring the government to increase the sum of the work plans from 150 pesos to 300 pesos ($50–100), and resisting the elimination of recipients. They look to extending their organization to include temporary workers and to joining forces with unionized workers to press for 'real jobs,' that is full-time, unionized, well-paid employment.

While the MTD-AV's assessment of the degree of support Kirchner enjoyed during most of 2003–4 was realistic, they failed to notice the decline of popularity and growing discontent by the end of the first year. Their perception of his 'progressiveness' was exaggerated and based on their favorable relation with his regime. They emphasized his 'resistance' to the IMF, but conceded under questioning that Kirchner abided with the IMF's conditions, continued to pay the foreign debt to IFIs and proposed to pay private debt holders at least in part. Moreover, MTD-AV conceded that poverty levels and inequalities remained unchanged and the privatized strategic industries and banks remained in the hands of foreign MNCs. The MTD-AV were willing to trade off immediate regime concessions in exchange for forsaking demands for structural changes, at least in the context of what they perceived to be an ebbing movement and a 'popular' regime.

In contrast, the Teresa Rodriguez Movement, which is opposed to the Kirchner regime, was deeply immersed in an internal debate about its future course. Roberto Martino, a leader of MTR, describes the conservative nature of Kirchner's regime: 'Cuts in labor plans, the reemergence of client politics, the breakup of the middle class–unemployed workers' alliance, no plans to reopen closed factories, paralysis of social movements, high levels of poverty, support for free trade agreement with the US and the European Union, and support for privatized industries—especially lucrative oil, electricity, and energy sectors—which could finance jobs and social services. In the eyes of the leadership of the opposition *piquetero* movements, Kirchner has 'recomposed' the national bourgeoisie, providing them with leadership and direction, while securing middle-class and even some transient popular support, even among the unemployed. They cite the support that Kirchner has secured from Hebe Bonafini of the human rights group *Madres de la Plaza de Mayo*, who has been a powerful ethical force, as providing legitimacy and reinforcing Kirchner's rule. The opposition *piqueteros* point to the middle class's 'increased repudiation of the *piquetero* road blockages and abandonment of neighborhood assemblies, because of their expectations that Kirchner will solve their problems.' In a word, the *piquetero* opposition describes Kirchner as being successful in the short term in reinforcing 'institutional politics,' partially channeling politics from the streets to the Congress and administration—and thus weakening the innovative assembly-style democracy which emerged before and subsequent to the December 20–21, 2001 uprising.

Among the opposition *piqueteros* there are two lines of thought and action. There are those who view Kirchner as merely a continuation of the old politicians (Polo Obrero-Trotskyist) and thus practice the same style of street politics, and those who believe that new times require new tactics and strategies. The latter position articulated by Martino criticizes those *piqueteros* 'who keep thinking it is 20/21 December 2001 and don't abandon road blockages.' He argues for 'new methods of struggle, for legitimation.' He maintains that most of the current protests are 'testimonial,' since the regime knows the protestors are isolated and that sooner or later the protestors will return to their homes. He also argues that the movement cannot just struggle for work plans and calls for a shift toward demanding a 'universal salary for all sectors of the working class (employed, unemployed) based on genuine jobs (stable full time employment).' This is to counter the influential regime propaganda that the unemployed who demand work plans 'don't want to work.'

This critical section of the *piqueteros* has no allies among the trade union confederation, and is in conflict with other critical *piqueteros* who continue to call for a 'general mobilization' leading to a 'general strike.'

Faced with general weakness in challenging state power in present circumstances, the MTR calls for engagement in 'territorial politics to encourage the masses to engage in local politics.' Martino argues for a two-stage process of first building municipal power to later gain access to national power and productive plans in order to recover the 'culture of work.' He calls for 'local self-administration and education to prepare workers for self-management.' What Martino calls for is a 'new state within the new state.' In order to move in this new direction the MTR propose to broaden the *piquetero* movement to become more inclusive to include employed workers, teachers, health workers, and other poorly paid temporary working sectors.

Work plans, which began as legitimate demands around which to organize group support based on assembly-style local self-governance, have in some instances turned into a tool for personal patronage of local leaders linked to the regime. Ironically, the system of local personal patronage relations has been justified by referring to 'horizontal structures,' an ideology popularized by the 'anti-power' ideologues.

The failure of the 'horizontalist' to achieve democratic control is in large part a result of the lack of class consciousness ('a class *for* itself'), which is a necessary development to exercise democratic control. Democracy in the *piquetero* movement without class consciousness did not lead to a sustained assembly-style political process. Instead, the popular rebellions and initial militancy led to a narrow focus on immediate consumption, social dependence on local *piquetero* leaders, and in some cases to political bosses.

The emphasis on 'autonomy' and 'spontaneity' of the *piqueteros* by the anti-power ideologues at the time of the rebellion was the other side of the coin to the subordination of the *piqueteros* to the new local regime bosses in its aftermath. Both phases reflect the absence of organized class-conscious political education. The absence of any strategic plan of action led to the dispersion of the movements toward a variety of reformist, collaborationist, and sectarian politics.

The *piquetero* movement was constituted on the basis of many workers who never were employed in factories and thus had little or no sense of proletarian class consciousness and among older workers who had been displaced from factory production for the better part of a decade. In many cases this led to 'individualistic solutions' rather than collectivist consensus subsequent to the initial rebellion.

The great accomplishment of the *piquetero* movement was the organization of the mass of unemployed for collective action. Its limitation is the failure to advance class consciousness, thus creating the current impasse and fertile terrain for the re-emergence of clientele politics under the 'benign' reign of the Kirchner regime. An emphasis on municipal rather than national issues fragmented the movement into hundreds of competing groups.

While the struggle for 'work plans' was initially an important first step to ameliorate hunger and infant malnutrition, the subsequent exclusive concentration on this issue has several negative consequences. In the first instance it created an '*assistencialista*' (social work) outlook among *piqueteros*—a dependence on minimum state transfers rather than a deeper questioning of the class nature of the state. The movements turned to militant struggles and confrontations (with street blockades, office takeovers) but in pursuit of narrow goals. With the establishment of the state-funded work plans, the *piquetero* movements became, in the words of Martino, a 'functional organization of the state, we became a social extension of the state—distributing the dole.' The rapid change in the means and goals of the *piquetero* movement needs serious critical reflections—not merely calls to 'return to the streets.'

In the current impasse between *piqueteros* engaged in isolated direct action and collaborationists supporting a 'moderate' neoliberal regime, a number of alternatives have been proposed. MTR-AV proposes struggling to pressure big enterprises to finance production projects of the *piqueteros*—so uniting employed and unemployed workers to promote jobs, class consciousness, and solidarity. This involves self-exploitation: 'voluntary work to obtain liberation,' according to Martino. The experience in Mosconi, a petroleum town where *piqueteros* have been able to extract resources form the privatized foreign-owned petroleum companies by blocking transport, is cited as an example. Mosconi secured legitimacy by supporting wages and security of employed workers and social needs of the community thus achieving ideological hegemony, a necessary prelude to challenging for state power. The problem with citing Mosconi is that it is vastly different from Buenos Aires and other metropolitan areas. Most of the unemployed were former petrol workers, with social, family, and union ties to the employed workers in a one-industry town. This is not the case with most industries in Buenos Aires. Moreover, the employed unionized workers in large-scale factories in Buenos Aires have shown little inclination to join with *piquetero* struggles, let alone support demands that the companies fund *piquetero* projects over increased wages. The more promising strategy is to join with low-paid public sector workers in joint strikes and combine demands for jobs and better pay.

THE FACTORY TAKEOVER MOVEMENT

Factory takeovers by unemployed workers reached a peak between 2001 and 2002 with over 10,000 workers operating over 100 enterprises. That movement is all but over. The political impetus for the takeover declined. Work plans absorbed some fired workers; and the Dulhalde regime via its judicial apparatus violently dislodged workers from the factories. Under Kirchner, the regime intervened and convinced the workers to convert the firms into profit-oriented cooperatives in exchange for legal recognition. Most adapted. Many 'occupied' firms now function as subcontractors for private firms employing onerous work conditions, under the tutelage of the state. They have to meet debt payments incurred by the previous owner and/or repay loans to the state or private banks. Most have lost their political cutting edge. They no longer act as part of a movement or see themselves as part of the class struggle. The workers who joined the takeover were acting merely to protect their jobs with little broader class consciousness. The leftist lawyers and activists in the solidarity movements did little or nothing to raise political consciousness, largely counting on the takeover itself (the factory occupation) to create class consciousness. Most of those leftists have been marginalized in the cooperatives. The major and significant exception is Zanon, the large, self-managed ceramic factory in Neuquen province. While many of the other worker-based cooperatives continue to operate and provide jobs, none of them retains the degree of worker management and control that remain hallmarks of Zanon. While many of the other factories which subcontract work extra hours at reduced pay to satisfy the price demands of their contractors, Zanon have taken on 140 new workers in their productive unit, increased production, improved quality, and maintained an egalitarian pay structure between skilled and unskilled workers. Unlike other factories taken over by workers and now converted into 'cooperatives,' the workers in Zanon through sustained class struggle prior, during, and after the occupation and political education have a high level of class consciousness. The Zanon workers are a leading force in promoting the semi-weekly newspaper *Nuestra Lucha* (Our Struggle) and have established firm ongoing relations and mutual support with the neighboring MTD.

The Kirchner government has, as of May 2004, refused to recognize Zanon as a worker-owned factory, despite the compliance of the Zanon leadership with all of the legal forms required by the regime to be classified as a 'cooperative.' While the Labor Ministry had promised to take up the matter—for over a year—the judicial system has once again taken the side of the bankrupt and corrupt ex-employer, and threatens to issue a judicial order to forcibly evict the workers. Former union bosses who were voted out of office, representatives of the World Bank, and members of the judiciary have all backed the employers.

Zanon raises a basic question. Why is their worker cooperative the only one in the country that the Kirchner regime has refused to recognize? We think the answer is to be found in the fact that in Zanon the state tutelage and paternalistic control which is exercised over the other factories will be hard to impose given its class-conscious leaders and members. Kirchner's functionaries operate like the old-style populist Peronists who observe the *forms* of worker representation in factories, while in practice manage control over the workers in accordance with the logic of the capitalist market. The danger to Zanon is real because the national network of solidarity, which sustained the movement, has in part unraveled. The *Madres* have embraced Kirchner as one of their own and no longer make their university premises available as a meeting ground for Zanon and its Buenos Aires supporters; the relaunching of their newspaper has encountered less than enthusiastic support from the declining and divided *piquetero* movements; and the intellectuals have returned to their academic duties or more 'current controversies.' While Zanon still remains a symbol of a successful alternative to capitalist management, it is no longer seen as a model to be followed by most unemployed or employed workers, who have signed up for work plans or are pressing for simple wage increases.

CONCLUSION: SHORT-TERM CONSOLIDATION, MEDIUM-TERM CRISIS

There is no doubt that President Kirchner has succeeded in consolidating support for his regime, making just enough personnel changes in the military, judiciary, and police to legitimize failed state institutions. He has acted with great shrewdness in meeting IMF conditions on budget and fiscal matters while striking a nationalist posture in resisting exorbitant demands to increase the fiscal surplus beyond 3 per cent and increasing payments to private bondholders. Most importantly, Kirchner has divided the social movements, co-opted many key trade union leaders, pensioners, and human rights leaders through minimum labor plans, some wage concessions, pension increases, and by ending the amnesty for military officials accused of human rights crimes. In May 2004 he announced a $185 million increase in pensions for 1.7 million retirees in the lowest bracket and a $35 million pay package for public sector employees who had lost ground under his wage restraint policies. In this case Kirchner was responding to strike action by the State Public Employees (ATE), backed by a threat from the Confederation of Argentine Workers (CTA) of a general mobilization.

The Argentine economy has capitalized on exceptional prices for its principal exports and improved taxation (up from 30 per cent in 2004 over 2003) to reap a record $3.9 billion. Faced with high growth in manufacturing, trade, and tax revenues the Kirchner regime has been able to placate middle-class consumers with cheap imports, encourage expectations among millions of unemployed workers with several thousand new job openings, and secure the acquiescence of several important *piquetero* leaders.

The time of popular rebellion against the ruling political class has temporarily passed. By the middle of 2004 several new sets of contradictions were emerging concerning Kirchner's macroeconomic policies. The organized workers and employees are demanding substantial wage increases to overcome losses from frozen reduced salaries; households are protesting Kirchner's granting rate increases to private foreign-owned power and energy corporations; Kirchner's continued support for foreign-owned (mainly Spanish) petrol and energy companies has led to a major energy shortfall, a partial shutdown of factories, and sharp increases in rates to consumers. Caught between his neoliberal pro-foreign capital commitments and the growing popular outcry at the unscrupulous price gouging of these same companies, Kirchner is facing his moment of truth. In April 2004, industrial activity declined by 4 per cent over the previous year due to energy shortages. Thousands have been fired.

As some of the more astute *piquetero* leaders recognize, the political conjuncture has changed, and yet the movements have not prepared for it either politically or organizationally.

What emerges from the extended and massive popular rebellion is that spontaneous uprisings are not a substitute for political power. Too many academics and political commentators failed to probe deeply into the inner strengths *and* weaknesses of the impressive but momentary social solidarity. There was little in the way of class solidarity that extended beyond the *barrio*; and the left parties and local leaders did little to encourage mass class action beyond the limited boundaries of geography and their own organization. Even within the organizations, the ideological leaders rose to the top, not as organized expressions of a class-conscious base, but because of their negotiating capacity in securing work plans or skill in organizing. The sudden shift in loyalties of many of the unemployed—not to speak of the impoverished lower middle classes—reflected the shallowness of class politics. The leaders of the *piqueteros* rode the wave of mass discontent; they lived with illusions of St Petersburg, October 1917, without recognizing that there were no worker soviets with class-conscious workers. The crowds came and many left when minimum concessions were made in the form of work plans, small increases, and promises of more and better jobs.

The process of movement domestication is located in a number of regime strategies executed in a timely and direct style. Kirchner engaged in numerous face-to-face discussions with popular leaders. He made sure the best work plans went to those who collaborated, while making minimal offers to those who remained intransigent. He struck an independent posture in relation to the most outrageous IMF demands while conceding on key reactionary structural changes imposed by his predecessors, namely the lucrative privatized ex-public firms. Lacking an overall strategy and conception of an alternative socialist society, the mass of the movement was easily manipulated into accepting microeconomic changes to ameliorate the worst effects of poverty and unemployment, without changing the structures of ownership, income, and economic power of bankers, agro-exporters or energy monopolies.

The question of *state power* was never raised in a serious context. It became a declaratory text raised by sectarian leftist groups who proceeded to undermine the organizational context in which challenge for state power would be meaningful. They were aided and abetted by a small but vocal sect of ideologues, who made a virtue of the political limitations of some of the unemployed by preaching the doctrine of 'anti-power' or 'no-power'—an obtuse mélange of misunderstandings of politics, economics, and social power (Besayag and Sztulwark, 2000; Colectivo Situaciones, 2001; Negri, 2001; Holloway, 2002). For the rest, the emergent leaders of the *piqueteros*, engaged in valiant efforts to raise mass awareness of the virtue of extra-parliamentary action and of the vices of the political class, were not able to create an alternate base of institutional power that unified local movements into a central force to challenge the state.

What clearly was lacking was a unified *political organization* (party, movement, or combination of both) with roots in the popular neighborhoods that was capable of creating representative organs to promote class consciousness and point toward taking state power. As massive and sustained as was the initial rebellious period (December 2001-July 2002) no such political party or movement emerged. Instead, we have a multiplicity of localized groups with different agendas which soon fell to quarreling over an elusive 'hegemony,' driving millions of possible supporters toward local face-to-face groups devoid of any political perspective.

Clearly, the slogan *'Que se vayan todos'* (Away with all politicians), which circulated widely among those recently engaged in struggle, turned out to be counterproductive as it further delayed or shortcircuited the necessary political education which an emerging political leadership required to deepen long-term mass engagement in revolutionary politics. Nonetheless the uprising of December 20/21, 2001 stands as a historic reference point for future struggles and a warning to US imperialism, the IMF, and the local ruling class that there are limits to exploitation and pillage. Moreover, the methods of extra-parliamentary action clearly were superior in ousting corrupt and abusive rulers to the electoral parliamentary–judicial processes.

By the end of Kirchner's first year in office (May 2004) the *piquetero* movement had re-emerged as the main opposition. In early May 2004 the mass movement returned with new allies, employed unionized workers, and new programmatic demands. These mobilizations included the blockade of 148 roads, highways, and bridges throughout the country with over 80,000 demonstrators. The principal demands centered on more work plans, an increase in the subsidies from 150 pesos (US$50) to 350 pesos (US$117) and opposition to state control and distribution of the work plans to the unemployed workers. Together with public employees, they called for increases in salaries for all state and private workers and an increase in pension payments for retirees. Equally important, all the *piquetero* groups were protesting judicial rulings outlawing street blockades. Road blockades were accompanied by a march through the city and protests at the headquarters of Repsol-YPF (the Spanish multinational petroleum company which bought the former state petroleum company) expressing popular repudiation for the increases in fuel prices and demanding a 'social price' for a tank of household cooking gas. The marchers also demonstrated outside the federal courts, protesting the court ruling condemning road blockades. Apart from Buenos Aires, road blockades took place in the provincial capitals of Jujuy, Salta, Tucuman, Santa Fe, San Juan, Mendoza, Chaco, Entre Rios, Corrientes, Misiones, Chabut, and Rio Negro. The day of struggle in early May 2004 was the first since the courts outlawed road blockades. An order by the Buenos Aires Minister of Security to 'clear the roads' was not implemented given the size of the demonstration. Many *piquetero* movements participated in the demonstrations, including *the Bloque Piquetero Nacional* (National Picketeer Bloc), led by the Workers Pole

(Trotskyists), the *Movimiento Independiente de Jubilados y Desocupados* (The Independent Movement of Unemployed and Pensioners) led by Raul Castello, and the *Corriente Clasista y Combativa* (Combative Class Current) former critical supporters of the Kirchner regime. In addition, the state employees' unions joined the demonstrations and road blockades despite obstructionist efforts by the pro-Kirchner trade union bureaucracy. During the third week of May 2004, the subway unions convoked a national assembly to launch a movement for a six-hour working day—to create jobs for the unemployed.

These renewed protests launched jointly by unemployed and employed workers, however, are offset by the decline of autonomy and increased vulnerability of the 'occupied factories.' Two symbolically important factories, Brukman (a clothing manufacturing plant) and Grissinopoli (a breadstick producer), passed from workers' control to state management, while the worker-run ceramic factory Zanon faces an imminent police-enforced eviction order. The *piquetero* movement, despite continued organizational divisions, still demonstrates a strong capacity to mobilize and convoke tens of thousands of militants on the basis of tactical alliances. The renewed activity is linked to *piquetero* organizations, which maintain an independent class perspective in relation to the Kirchner regime. Those groups that have taken a position of collaboration, critical or not, have become enmeshed with the state and have become politically and organizationally incapable of responding to the rising discontent among unemployed and poorly paid employed workers.

The first ten months of Kirchner's reign was a period of high expectations among the populace. The hope that better days were ahead has worn off. The subsequent US$50 rise in monthly salaries for the lowest paid public employees and pensioners still falls short of rising fuel, energy and electrical prices, which Kirchner has generously granted to the privatized MNCs.

Kirchner's pursuit of 'normal' national capitalist development has revealed its structural weakness in the face of the energy, gas, and electrical crises provoked by the foreign-owned MNCs. During years of exorbitant profits the MNCs made little or no investments in new oil pipelines, infrastructure, or exploration to meet rising demands. In the present period they have lowered output in the generator plants by closing them for 'maintenance.' The MNCs created an artificially acute 'shortage' of energy, blaming government regulations. Moreover, the original privatization contracts gave them 54 per cent retention of petrol and gas to dispose as they wish, resulting in lucrative overseas export deals with Chile, Uruguay, and Brazil, but further limiting supply to Argentine industrial and household consumers. Faced with corporate blackmail and after engaging in a bit of populist rhetorical demagogy by criticizing the MNCs, Kirchner caved in and granted the price increases. In an apparent move to pacify nationalist opinion, Kirchner promised to launch a state petroleum company, which would construct the infrastructure to facilitate the private exploitation and commercialization of energy and petroleum.

Kirchner's 'populist theatrics' have less effect over time. The masses of consumers directly experience a decline in real income and a sudden and steep increase in prices. The net result of the MNC-induced 'energy crisis' is layoffs and plant shutdowns (increasing unemployment and lowering wages) while increasing the number of impoverished households, who literally live in the cold. While the collaborationist 'pragmatic' *piquetero* organizations gained short-term and limited favors (more work plans, local appointments, small-scale financing) Kirchner's embrace of the privatized monopolies, continued debt payments, and restrictive budgeting policies has prejudiced the poor over the medium term. As a result, the axis of *piquetero* politics has moved from the 'pragmatic collaborationist' leaders incapable of responding to the energy and income crisis to the more militant class-oriented *piquetero* leaders and organizations. Foreign capital located in the strategic sectors of the economy dictates the costly terms under which national capital function. The hard currency earnings to finance national capital depend on volatile commodity prices. Both structural factors inhibit any possibility of sustained national capitalist growth. Add to that the high propensity of 'national capital' to send their profits abroad and to invest in speculative activity in Argentina and it is easy to understand the resurgence of an Argentina crisis.

The first wave of mass mobilization, roughly from January to July 2002, generated the mass *piquetero* movement and the capacity of the unemployed workers to engage in mass direct action. This created a degree of class consciousness among hundreds of thousands of activists in the poorest *barrios*. The decline of the movements (August 2002 to May 2003) coincided with the regime-sponsored $50 work plan, internal strife among the *piquetero* groups, and the hopes of an electoral solution via Kirchner. The retreat deepened during the first year of Kirchner's term of office (June 2003—April 2004) as he successfully co-opted a substantial proportion of *piquetero* leaders through their incorporation in the state apparatus and financing of small-scale projects and symbolic gestures. However, the initial enthusiasm for Kirchner is giving way to strikes and protests. The impoverished workers realize that work plans have not led to real jobs with liveable wages; they know that local projects do not solve the problems of low wages, rising prices, and malnourished children. Discontent began to surface in early March 2004 as small contingents of workers resumed road blockades and large-scale confrontations took place in the provinces between corrupt authoritarian pro-Kirchner governors and public employees, the unemployed, and human rights supporters. By May 2004 dissatisfaction over Kirchner's energy price hikes, frozen salaries, and 20 per cent disguised (work plans) and open unemployment erupted in organized street action.

The key factor is the temporary and fragile convergence of demands between the low-paid public and privately employed workers, energy consumers, and the unemployed. The defection of the 'middle class' to the right (once they had recovered their savings), which led to the temporary isolation of the *piqueteros*, could be compensated if a new coalition of unionized workers and the unemployed gained a firm footing. No doubt Kirchner will make some concessions to divide this burgeoning coalition, especially via the collaborator trade union bosses. However, his margin to 'divide and rule' is limited by the end of the favorable international prices for Argentina exports. He can no longer count on the support of workers with *future* expectations of jobs and increases in the standard of living—the 'future' has arrived. Current realities no longer convince the 50 per cent still living in poverty. With the prices of mineral (except petrol) and agrarian exports declining Kirchner lacks the margin to pay the debt, raise wages, and create jobs. Thirdly, he has demonstrated his commitment to sacrificing local living standards and growth to meet the profit demands of the energy MNCs. Finally, Kirchner's cynical play on 'opposing' the IMF and then transferring billions in debt re-payment to it is highly unlikely to continue to deceive the majority of Argentines.

It remains to be seen whether the militant *piquetero* movements can build durable alliances with employed workers, deepen the class consciousness of its activists, and create a broad-based political movement that unites the still deeply divided movement.

As the current mini-boom ended in April/May 2004, as Argentine agro-export prices declined, Kirchner had neither the economic resources nor the ideology to sustain his balancing act. In addition, his refusal to free 4,000 popular activists from judicial proceedings—some facing five to ten years' imprisonment for political offences—stands in the way of any lasting alliance. The resurgence of the mass class struggle in Argentina profits from its successes through direct action. It is in a position to confront structural problems (poverty, unemployment, low salaries) and learn from its limitations—the absence of a mass-based national political party movement aiming at state power and the resocialization of strategic sectors of the economy. Once again, the mass movement will learn that none of the basic problems will be solved by an 'alliance' with the national bourgeoisie, even a benign Kirchner version.

3
Lula and the Dynamics
of a Neoliberal Regime

The election of Luis Inacio ('Lula') da Silva raised great expectations among the center-left. For most leftist writers, his election heralded a new epoch of progressive change, which, while not revolutionary, defined the 'end of neoliberalism.' Noted progressive religious figures, like Leonardo Boff, announced that imminent change would challenge US hegemony and lead to great popular participation. Frei Betto, a close associate of Lula, launched a vitriolic attack on critics who questioned some of Lula's appointments, citing his popular roots as a former metal worker and union leader a quarter of a century earlier. Leftwing members of the Workers Party, Olivo Dutra and Tarso Genero, appointed to minor ministerial positions in Lula's cabinet, called for the 'disciplining' (i.e. expelling or silencing) of a dissident Workers Party senator, Heloisa Helena, who objected to the PT's support for rightwing Senator Jose Sarney as president of the Senate. European, US, and Latin American progressives and leftists and their movements, NGOs, parties, and journals joined the celebration of the Lula presidency, his 'progressive agenda,' and his 'leadership in the fight against neoliberalism and globalization.' While over 100,000 at the World Social Forum in January 2003 at Porto Alegre cheered Lula as a hero of the left and precursor of a new wave of leftist regimes (along with Lucio Gutiérrez and Hugo Chavez of Venezuela), one of Lula's intellectual supporters (Emir Sader) pleaded with Lula not to go to Davos to plead his case for foreign investment to the world's most rapacious speculators and richest investors.

In addition to the great majority of the left intellectuals, NGOers and politicians who aggressively and unquestionably support Lula as a new progressive force, the Brazilian and foreign financial media, international financial institutions (IMF, World Bank, Wall Street, City of London, and prominent rightwing leaders like British Prime Minister Tony Blair and US President Bush) praised Lula as a statesman and 'pragmatic leader.' In other words, big business, bankers, and rightwing political leaders see Lula as an ally in defense of their interests against the left and the mass popular movements.

In this chapter we analyze and evaluate the expectations of the left and capitalist perceptions in light of political and economic realities. Our approach is based on what we might term the methodology of class and regime analysis. This methodology is based on the following procedures: an examination of (1) the historical dynamics of the Workers Party (PT) in the pursuit of political power; (2) the relation between the PT and national, state, and local governments where it has held power; (3) Lula's electoral campaign and the political alliances and pacts that accompanied it; (4) the nature of Lula's 'economic team' in terms of the identity, background, and politico-economic practice of the key ministerial and economic functionaries; (5) the macroeconomic and macrosocial performance of the regime in its first year of office and its social impacts, viz. the beneficiaries and losers of the regime's economic policies sector by sector and by class; and (6) the political and social dynamics of the regime's second year in terms of the transformation brought about in the society's political economy.

YEAR ZERO: THE HISTORICAL DYNAMICS OF THE WORKERS PARTY (PT)

The PT's publicists refer to it as a workers' party, with reference to its supposed ties to social movements and its deep involvement in class and other social struggles. This was the case at its founding over two decades past. But the most significant fact about the PT is its qualitative change over the last quarter of a century. Several fundamental changes have taken place in the PT: (i) its relation to the social movements and their struggles; (ii) the internal structure of the party and the composition of the delegates to its congress; (iii) its program and political alliances; and (iv) its style of leadership.

At its very foundation, the PT was a party with a strong component of social movements—landless workers, urban *favelados* (slum-dwellers), ecologists, feminists, cultural and artistic groups, progressive religious and human rights activists, and the major new trade unions, including the metal workers, teachers, banking, and public employees. The PT grew in membership and influence from its direct involvement in the movement struggles. The electoral campaigns largely complemented the extra-parliamentary struggles in the beginning. Over time and with greater electoral successes, the 'electoral' sector of the PT gained control of the party and slowly redefined its role as basically an electoral apparatus, paying lip service to the social struggle and concentrating its efforts inside the apparatus and institutions of the state, forming *de facto* alliances with the bourgeois parties. A minority of the 'electoral party,' the left wing, continued to support the social movements—from the institutions—providing legal defense, denouncing state repression, and giving verbal encouragement at mass gatherings. What is clear, however, is that all tendencies of the electoral party, left, center, and right, were no longer engaged in the day-to-day mass organizing, except prior to election campaigns.

The second basic change was in the composition of the party and the party congresses. By the mid-1990s the great majority of the party apparatus was made up of full-time functionaries, professionals, lawyers, public employees, university professors, and other middle- and lower-middle-class employees. The 'voluntary activists' disappeared and/or were marginalized as the party turned from mass struggles to office-seeking and wheeling and dealing with business groups and a diverse array of center-left to center-right parties.

The last congress of the PT prior to Lula's election was overwhelmingly (75 per cent) middle-class, mostly functionaries, with a sprinkling of trade union, MST, and human rights leaders. Clearly, the PT was no longer a 'workers' party,' whether in its composition, its delegate congress, or in its relation to the social movements prior to the elections. Moreover, many of the elected officials of the PT at the municipal and state level were engaged in the same kind of cross-class alliances with business groups and bourgeois parties that the PT would follow in the presidential campaign of 2002 and once in power. In other words, the right turn of the PT at the national level was preceded by a similar pattern at the state and municipal levels during the 1990s. More significantly, many of the key party leaders and subsequent advisors to Lula were already practicing neoliberal office holders, even as the national party program spoke of socialism, anti-imperialism, and repudiation of the foreign debt. As the 2002 elections approached, the national leadership of the PT, with Lula leading the way, eliminated all the programmatic references to socialism and anti-imperialism, in line with the practices of the neoliberal office-holders in the party and with the majoritarian support of the middle-class party delegates.

The third significant change in the PT is the evolution of its program. Essentially, the programmatic changes took place in four stages. First, during the 1980s, the PT stood for a socialist society based on assembly-style democracy, linked to the social movements. The PT called for a repudiation of the foreign debt, a sweeping land redistribution with state financial, technical, and marketing support, socialization of banking and foreign trade, and national industrialization (with some sectors calling for the expropriation of large industries and others for worker co-management). These radical positions were debated openly and freely by all the tendencies (from Marxists to social democrats), who even published their own newspapers and dissent.

Beginning in the late 1980s and into the late 1990s, the PT moved to the right, the axis of power shifted toward a 'social-democratic position' (support for a welfare state) while the Marxist-left continued as a strong minority tendency. This was the first major change in the PT's program, leading to a second stage in its politics. The social democrats controlled the increasingly middle-class party apparatus, while the Marxists organized their opposition from within the same apparatus, few, if any, turning to mass organization to counter their growing weakness in the party machinery. While the formal program retained the earlier radical demands, in practice most of the newly elected governors and mayors did not challenge existing property relations. The radical wing of the elected officials in Porto Alegre introduced the notion of a 'participatory budget,' involving neighborhood committees, but failed to municipalize any essential services, including transport, or encourage land occupations or the demands of landless workers. Moreover, the participatory budget was based on the funds allocated by state and municipal regimes, which established the overall budget priorities. Politically, this meant that even the radical PT learned to coexist and cooperate with the established banking, industrial, and real estate elites so that the debate between the minority Marxist and dominant social-democratic wings of the PT was over programmatic language, while the differences of practice between them were in fact quite narrow.

The third phase of the PT, roughly between the end of the 1990s and the run-up to the elections, saw a further shift to the right in programmatic terms. Even the rhetorical references to Marxism, socialism, and foreign debt repudiation disappeared. The party leadership was in full transition to social liberalism, combining anti-poverty populist rhetoric with the pursuit of alliances with neoliberal business, banking, and agro-export elites. During the election campaign, Lula repudiated a referendum on ALCA organized by the MST, sectors of the progressive church, and other leftist groups. Instead, the PT called for 'negotiations' to improve ACLA. The PT embraced a pact (in June 2002) with the IMF and acceded to its dictates on fiscal austerity, a budget surplus to pay bondholders, cuts in public spending, and respect for all privatized enterprises. The social aspects of this liberal program were the declaration in favor of a gradual agrarian reform (of unspecified dimensions), a 'zero-poverty' agenda, providing family food subsidies, and land titles for urban squatters.

The final phase in the evolution of the PT's program began in 2003 as a presidential party. The PT government embraced an orthodox neoliberal program. Despite promises of increased social spending, the Lula regime slashed budgets, imposed fiscal austerity, raised interest rates to attract speculative capital, and negotiated with the US to lower its trade barriers. In other words, for the Lula regime its differences with the US concern converting Washington to a consequential free market economy. Most of the leftists around the world who see the victory of the PT and Lula as the advent of basic or at least important social changes benefiting the poor and redistributing wealth and land, base their views on long outdated images of reality. Over the past few years the militants who built the party through grass-roots movements have been replaced by 'neo-Lulistas,' upwardly-mobile functionaries, professionals with no history of class politics, who have joined the party to secure the perks of office and oil the wheels of business liaisons. What remains of the older reform social democrats have been shunted to marginal ministries or, if they dare to question the neo-Lulista hegemony, are subject to punitive measures for 'violating party discipline,' including expulsions.

As in the case of Britain where Tony Blair's neoliberal pro-imperialist 'New Labour' replaced the traditional social democratic Labour Party, likewise Lula's orthodox neoliberal strategists have created a 'New Workers Party' without social content, without democracy.

LEADERSHIP AND PARTY DEMOCRACY

From its founding to the late 1980s, the PT had a vibrant, open, freewheeling internal life. Members came to general assemblies and debated with leaders and held them responsible for their policies, speeches, and presence (or non-presence) at popular demonstrations. Leadership was collective and the different political tendencies argued their positions without fear of expulsion or discipline. To outside observers, particularly conventional US social scientists, the internal party life was 'chaotic.' Yet great advances were made in recruiting new activists, militants volunteered for political activities and electoral campaigns, and the party advanced despite the universal hostility of the mass media.

By the end of the 1980s, however, the social-democratic electoral wing of the party was in the ascendancy and proceeded to discipline and expel some sectors of the radical left of the party. Assemblies were replaced by leadership meetings of full-time functionaries who implemented policies and then opened the floor to debate with their radical counterparts in the party apparatus. Thousands of activists began to drift away, in part by the growth of clientelism, in part by the emerging vertical structures, and in large part because the party turned almost exclusively toward electoral politics. Most outside observers continued to write about the PT as if it was still the 'horizontal grass-roots' organization of earlier years, confusing the debates between the different tendencies (left, right, and center) of the party apparatus with the earlier popular assemblies. By the election of 1994 and continuing with greater intensity thereafter, the PT became a personalist party organized around Lula, as the embodiment of the popular will, and the competing party barons in their power bases in state and municipal governments. Increasingly, voluntary party activists were replaced by paid functionaries, political appointees to public office and public relations specialists in polling, image-making and television ads. Strict rules on electoral financing were breached as the leadership sought and accepted funds from state contractors to pay for the new and expensive mass media style of electoral campaigning.

With the new millennium, the party was run by a small nucleus of close advisors and a small elite of party bosses led by Ze Dirceu, who surrounded Lula and encouraged his personalist and increasingly authoritarian centralized leadership. Programs were no longer open to serious debate. The party program, everyone was told, was what Lula wanted in order to run for office, or later in order to win the campaign. Lula decided, with his coterie of advisors, to form an alliance with the rightwing Liberal Party without consulting anyone, let alone the mass base, concerning this strategic shift. In government he formed an alliance with the PMDB in a similar fashion. The same group pushed through a new social-liberal program via its control of the full-time functionaries at the party congress just prior to the 2002 elections. A top-down personal leadership became the hallmark of the PT, a far cry from its earlier horizontal structure.

The shift to authoritarian political structures facilitated the repudiation of all of the PT's remaining social reformist demands. As the traditional PT program was discarded and Lula's opening to the right deepened, his advisors increasingly projected the image of Lula as 'the man of the people,' the 'compassionate northeasterner,' the 'metal worker president.' Lula played the dual roles of neoliberal and 'worker president' to perfection: to the *favelados* he provided hugs, tears, handouts, and promises; to the IMF he guaranteed budget surpluses to pay bondholders, the firing of public sector employees, and the promotion of agro-export elites.

The PT is a party which aspires to represent an alliance between domestic big industrialists and agro-business interests and overseas bankers: it secures the loyalty of labor bureaucrats for 'social pacts' through lucrative pay-offs and 'pacts' which allow business to reorganize the workplace, fire workers with little or no severance pay, and increase part-time and short-term employment, in exchange for which trade union bosses will receive future government posts and monetary remuneration. The appointment of trade union bureaucrats and left PT members to the Agrarian Reform and Labor Ministries is designed to mollify the unions and the MST with symbolic, not substantive, representation. The job of the PT ministers is to preach 'patience' and to make inconsequential radical speeches at industrial workers' and landless workers' meetings. All the 'leftwing' ministers with limited budgets and under a pro-business economic strategy were totally incapable of pursuing any substantial reform programs. They pleaded with the dominant neoliberal economic ministers for residual financial outlays, an undertaking that rarely, if ever, succeeded. Eventually, the impotent leftist ministers were ousted and others adapted to the liberal orthodoxy and argued for what they will call 'new realism' or 'possibilism.'

The PT as a dynamic movement based on the working-class peasant party is dead. Long live the neo-Lulistas and their paternalistic leader!

THE ELECTORAL CAMPAIGN

The past weighed heavily on the mass vote in favor of Lula and the PT; the present and future, however, raise hopes and open new vistas for the overseas bankers and domestic elites. These two distinct and polarized perceptions and interests are important to keep in mind in analyzing Lula's electoral appeal among the mass of the poor and the pro-business economic policies that he promoted before and after his election. Lula's political agreements with the right and economic pacts with the IMF during his election campaign reflect the evolution of the PT over the previous decade and foreshadowed the orthodox neoliberal policies he took immediately upon assuming office.

Several key factors during the electoral campaign prefigured the neoliberal cabinet appointments and policies followed after the election: (1) Lula's economic and campaign advisors; (2) the choice of political allies; (3) the nature of the socioeconomic program; (4) the agreement with the IMF; and (5) promises to meet and reach agreements with US officials, overseas bankers and investors, and the domestic industrial and agro-export elites.

A small nucleus of campaign advisors played the major role in shaping Lula's presidential campaign, advisors who were long known for their neoliberal credentials. In effect, Lula bypassed all the democratic norms and party statutes in organizing his campaign, including the process of selecting his vice-presidential running mate and formulating his future program. Three advisors stand out: Antonio Palocci, the former PT mayor of Ribeiro Preto, a city in São Paulo state, who coordinated the PT's campaign platform and established solid links with the business elite. He was the PT's top spokesman on economic policy during the electoral campaign and headed the transitional team after the elections. Palocci also engineered the PT's agreement with the IMF and was the architect of the orthodox monetarist and fiscal austerity economic policies. Lula later appointed him Finance Minister. As mayor of Ribeirao Preto, Palocci allied himself with the local business elite and the sugar barons (*Financial Times*, November 15, 2002, p. 3). He privatized the municipal telephone and water companies and partially privatized the municipal transport system. Apart from some low-cost housing development, his neoliberal policies were uniformly detrimental to the poor. Crime rates increased, as did the queues at local hospitals. Seven years into his government only 17 per cent of the city's wastewater was being treated. Equally serious, water bills and regressive taxes increased and the public prosecutor investigated 30 charges of government corruption in relation to public works contracts. As a result of Palocci's reactionary policies, Lula barely won the popular vote in Ribeirao Preto (in contrast to his 24-point national margin), a result likely to be repeated in the next presidential election.

Jose Dirceu, former president of the PT, is Lula da Silva's most influential advisor for almost decade. He has been the major force in engineering the transition from social democracy to neoliberalism. He was appointed chief of the president's cabinet and presides over the everyday affairs of the president's agenda and appointments, as well as exercising disciplinary power over PT deputies and senators to ensure that they vote the neoliberal line on appointments, legislation, and priorities. Dirceu, known as the 'Commissar,' has already demonstrated his heavy hand in the expulsion of Senator Heloisa Helena for refusing to vote in favor of former Bank of Boston CEO Henrique Meirelles as head of the Central Bank and rightwing Senator Jose Sarney as president of the Senate.

The third close advisor to Lula during the campaign was Marcos Lisboa, an orthodox liberal professor and staunch monetarist. According to the Brazilian daily, *Folha de São Paulo* (December 22, 2002), he was selected by Palocci to formulate Lula's economic strategy. He is part of a large group of neo-Lulistas who jumped on the presidential bandwagon in the last weeks of the presidential campaign when it was clear that Lula would win. This inner circle is backed by a wider ring of neoliberal senators, governors, and mayors who are deeply allied with business interests and who promote privatization policies.

Those key advisors, along with Lula, decided on the political alliances to promote Lula's election. The strategy was first to consolidate control over the PT to ensure big city support, concentrating power at the top and then moving to the neoliberal right to gain the support of the small towns and backward rural areas, and, more important, big business financing. Lula selected Alencar from the Liberal Party as his vice presidential partner. This won him support from a substantial minority of Brazilian business groups and among rightwing evangelical groups backing Alencar, himself one of the richest textile capitalists in the country and no friend of the trade unions, least of all those employed in his textile mills.

While the left PT objected verbally, they eventually swallowed Da Silva's decisions, since they had no recourse, no chance of changing the selection since the issue was never discussed outside of Lula's coterie. Dirceu, Palocci, and their regional party allies then proceeded to form political pacts with center-right and rightwing parties across the whole political map, in different states of the country. In some cases, the national leadership's pacts with the right undermined local PT candidates, leading to the loss of several governorships. What is clear from these electoral alliances with rightwing parties is that they were not 'opportunist' moves or merely electoral tactics. Rather, the alliances coincided with the neoliberal ideology within Lula's inner circle and among key sectors of the PT's congressional representatives. The new rightwing allies plus the recently recruited neo-Lulistas in the PT served as a counterweight against the left wing of the PT, further reducing their influence in the party and the government. This was evident with regard to two important developments during the campaign: the PT's program and its pact with the IMF.

Lula and his neoliberal team made a consistent and coherent effort to demonstrate their neoliberal credentials to several key groups, including Wall Street, the Bush administration, the IMF, and the principal Brazilian banking and industrial elites. Palocci was a key player in all of these key negotiations.

The electoral program of the PT spoke to all of the major concerns of the financial and industrial elites. Privatized enterprises would be respected. Foreign debt payments would continue. Tight fiscal policies would be rigidly adhered to. Labor and pension 'reform' would be at the top of the agenda (reform = weakened trade union rights and labor legislation, and reductions in public sector pensions). There would be no indexation of wages and salaries, but there would be for bonds and debt payments.

The PT's program was a clear continuation of outgoing President Cardoso's disastrous neoliberal policies and in some cases even a radicalization of his liberal agenda.

To demonstrate their liberal orthodoxy to the bankers and industrialists further Lula's team signed up to a pact with the IMF only a few weeks before his electoral victory. In exchange for securing a $30 billion loan over a four-year period, Lula agreed to a strict adherence to all the typical retrograde conditions set by the IMF. Once in office Lula even went beyond these harsh measures. The IMF agreement included the typical recessionary measures maintaining inflationary control by withholding large injections of fresh capital to stimulate growth, acquiescence in the disastrous privatization program introduced by outgoing President Cardoso, and a budget surplus target (beyond what is paid in interest payments) of 3.75 per cent of gross domestic product, thus guaranteeing in advance that little or no funds would be available for any of Lula promises of 'zero poverty,' let alone financing a comprehensive agrarian reform.

Lula appointed a former president of a US multinational investment bank (Fleet Boston Global Bank) Henrique Meirelles as the head of the Central Bank. Meirelles supported Cardoso's orthodox neoliberal agenda and admitted to voting for Jose Serra, Lula's opponent in the

259-7772

presidential election. The Finance Ministry is in the hands of orthodox neoliberal Antonio Palocci, a member of the extreme right wing of the PT. Luiz Fernando Furlan, millionaire chairman of the agricultural company Sadia, was appointed head of the Trade and Development Ministry. Robert Rodriguez, president of the Brazilian Agribusiness Association and strong advocate of genetically modified crops, was selected to be Agricultural Minister (*Financial Times*, December 17, 2002, p. 3). As a spokesman for the largest multinational commodity giants, Rodriguez joins Monsanto, the international agricultural and biotechnology group engaged in a longstanding battle to permit sales of GM Roundup Ready soya seeds. Lula's economic team of neoliberal ideologues and millionaires outlined the pro-big business agenda even before taking office. From the beginning, it was clear that popular expectations among the 52 million who voted for Lula and the 200,000 who cheered his inauguration would be profoundly disillusioned once his economic team began applying the IMF agenda. Lula further extended the reach of the right by reappointing Cardoso's supporter Gilberto Gil as Cultural Minister, PT former governor of Brasilia, Cristovan Buarque, a strong advocate of privatization, as Education Minister, and Cardoso's former ambassador to the US, Celso Amorin, as Foreign Minister.

To pacify the center-left of the PT, Lula appointed a number of officials to ministries that were largely impotent given the tight fiscal and monetary policies imposed by his big business economic team. By co-opting the left to the marginal ministries Lula deflected popular tensions and cultivated illusions among the leaders of the social movements that his is a 'balanced' regime. For the seven trade unionists, four women and two blacks in the cabinet, upward mobility outweighs concerns about neoliberal policies. By the end of the first year in office, having consolidated total political control over the PT and firmly established his neo-liberal agenda, Lula gave the boot to even his 'loyal' PT moderates, unceremoniously removing Cristovam Buarque from the Education Ministry, Benedita da Silva from the Social Promotion Ministry, and Jose Graziano (author of the Zero Hunger Program) as Minister of Food Security. He added two ministers from the rightwing Brazilian Democratic Movement Party (PMDB) to secure the legislative and political bases for pushing through his extreme 'neoliberal' agenda (*La Jornada*, January 24, 2004). To ensure that neoliberal policies are implemented, Lula is pushing a constitutional amendment that will make the Central

Bank more responsive to foreign investors and bankers by making it 'autonomous' of the national legislature and president.

Parallel to the selection of the big business cabinet, Lula's inner team of Paolucci, Dirceu and their economic advisors moved quickly to demonstrate their allegiance to US imperialism, the big investment houses, and the Brazilian industrial elite. Between Lula's election and his inauguration, neoliberal advisors assured the US that ALCA (Free Trade Area of the Americas) was a framework for negotiations. Three weeks after Lula's election, Peter Allgeier, deputy US trade representative, stated: 'We will be able to work with the new [da Silva] administration on trade issues across the board in the World Trade Organization, in the FTAA and bilaterally. I feel very positive after having spoken to a number of people associated with the upcoming president' (*Financial Times*, November 22, 2002, p. 4).

Immediately after being elected, the Da Silva team was already laying the groundwork for close economic ties to US imperialism, a point missed by many of the Brazilian left intellectuals like Emir Sader, who continued to praise Lula's 'nationalist' foreign policy (*Punto Final*, December 2002, p. 2). A few weeks before his inauguration Lula met with Bush in Washington, where the two leaders agreed to a trade summit for Spring 2003. In addition, he met with US trade representative Robert Zoellick to discuss how the co-chairs of the negotiations on the ALCA could expedite its implementation (*Financial Times*, January 22, 2003, p. 12). The PT regime's pro-ALCA, pro-US position was apparent when Lula refused to support the 2002 popular referendums on ALCA and the US base in Alcantara in Maranhao State, despite more than ten million participants. Lula's decision to repudiate the 95 per cent of voters who opposed ALCA and the US base, and to move toward greater subordination immediately after the elections, is indicative of the massive deception perpetrated by his electoral campaign. As the inauguration neared, the neoliberal nucleus running the government made it clear that budget austerity and high interest rates would take precedence over poverty reduction and development initiatives.

Though many on the left of the PT had doubts about Lula's alliance with the hard neoliberal right, including electoral pacts with ex-president Jose Sarney, and the corrupt ex-governor of São Paulo Orestes Quercia and Paulo Maluf, they continued to describe the Lula regime as a government 'in permanent dispute and tensions,' without a fixed direction. Blinded by the presence of former leftists in marginal cabinet posts, they overlooked the deep structural and policy ties of the key economic and foreign policy-makers. The only 'dispute' was between the foreign bankers and big industrialists over interest rates.

Lula pressed all the buttons to please Bush. He publicly criticized Presidents Chavez of Venezuela and Fidel Castro of Cuba prior to his inauguration. His inauguration speech was a masterpiece of duplicity, a double discourse to set his lower-class supporters dancing in the streets and assuring foreign bankers that his regime was their regime. Lula's speech spoke of 'changes,' 'new roads,' and the 'exhaustion of a [neoliberal] model,' which he then qualified by speaking of a 'gradual and continuous process' based on 'patience and perseverance.' He then spoke of 'zero hunger' as the priority of his government. He spoke of agrarian reform and developing the internal market, but he also came out in favor of agro-export elites and free trade and against protectionism and subsidies. Having appointed the most rigid neoliberals to every key economic post, he could not possibly be taking a 'new road.' After signing up to the IMF austerity budget there was no way he could finance new employment or 'zero hunger.' By setting anti-inflationary measures designed by the IMF as a priority there was no way Lula could promote the internal market.

The double discourse belied a single practice, to continue and deepen the model that he denounced as leading to stagnation and hunger. Once in office Lula very early on demonstrated the vacuity of his promises of social welfare.

Implementing neoliberalism: Lula in power, year zero

One thing must be said about da Silva's economic team: they lost no time in fulfilling their pre-inauguration promises to the IFIs, international bankers, and the local industrial elites. There is no balancing act (*Financial Times*, January 24, 2003, p. 2) between the 52 million voters with expectations of social improvement and da Silva's commitment to the economic elites. Few ex-left governments have moved so rapidly and decisively to embrace and implement a rightwing agenda as has this regime.

In line with meeting the demands of the IMF and the economic elites, the regime slashed the budget by $3.9 billion (*Financial Times*, February 11, 2003, p. 66; *La Jornada*, February 11, 2003). Included in the budget cuts were reductions in the promised minimum wage from $69 to $67 per month to take effect in May 2003, five months after taking office. Given inflation, this reduced the minimum wage below the miserable level of the Cardoso regime. Over $1.4 billion of the

$3.9 billion budget reduction came from the social budget. A closer analysis of the budget cuts reveals that the reductions affected food programs, education, social security, labor, agricultural development, cities, and social promotion. Altogether, social cuts came to 35.4 per cent of the budget reduction. Even Da Silva's much-publicized pet project, 'zero hunger,' was slashed by $10 million, leaving $492 million for meeting the needs of the 40 million malnourished Brazilians. The budget cuts mean the funds allocated for the hungry amount to $10 a year or $0.85 a month or 2.5 cents a day.

The major reason for the social and other budget cuts was to increase the budget surplus to meet IMF and debt repayments. Da Silva's neoliberal Talibans increased the surplus from 3.75 per cent of GNP agreed to with the IMF in June 2002 (under Cardoso) to 4.25 per cent in February 2003 under the leadership of the former metal worker and 'people's president.' In other words, da Silva increased the budget allocation to meet debt obligations from $17 billion to $19.4 billion, or by nearly 14 per cent. The $2.4 billion addition was a direct transfer from the social budget to the foreign and domestic bondholders. Da Silva thus transferred funds from the very poorest classes to the very rich.

Da Silva's budgetary policies have deepened Brazil's infamous inequalities not reduced them. His theatrical gestures of asking the poor who voted for him to 'pardon' him for prescribing this 'bitter medicine' will certainly not bring much sympathy from the millions of minimum wage workers who will see their meager incomes and social services decline. The cuts in government spending did not provide any stimulus to the economy but instead deepened the economic recession.

Da Silva's neoliberal appointees to strategic economic positions established the strategic economic framework for the formulation of macro- and microeconomic and social policy. To understand what has transpired since he took office it is essential to understand the underlying philosophy which guides his regime and to set aside his theatrical antics before the mass public and the populist gestures directed at pacifying the poor, the social movements, and dissident members of the PT.

The operating philosophy of the PT regime has several key postulates: (1) Brazil is in a crisis which can be addressed only by satisfying the austerity policies promoted by the international financial institutions in order to secure new flows of loans and foreign investment, which are identified as the principal vehicles for development (*Financial Times*, January 16, 2003, p. 2). (2) Brazil will grow only by providing incentives to domestic big business, agribusiness, and foreign multinationals (see Lula at Davos, *Financial Times*, January 27, 2003, p. 2). These incentives include lower taxes, reducing labor welfare provisions, and strengthening business positions in labor management negotiations. (3) The free market, with minimum state intervention, regulation, and control is essential for solving the problems of growth, unemployment, and inequalities. The major task set by Da Silva's economic team is to promote Brazilian exports to overseas markets over and against domestic markets and to pressure the US and Europe to liberalize their markets (*Financial Times*, January 16, 2003, p. 2). (4) Growth will eventually result from price stability, foreign capital flows, tight fiscal policy, and above all strict payment of public and foreign debts, hence the need to slash government budgets, particularly social budgets, to accumulate a budget surplus for debt payments and to control inflation. Once stability (the 'bitter medicine') is achieved, the economy will take off into market-driven export growth, financing the poverty programs to alleviate hunger. 'Premature' welfare spending, raising the minimum wage, extensive poverty programs, and agrarian reform would 'destabilize' the economy, undermine 'market confidence,' and lead to deepening the crisis and worsening the condition of the people (*Tiempos del Mundo*, Dominican Republic, February 20, 2003, p. 7).

These doctrinaire neoliberal philosophical assumptions of da Silva's economic policies provide the basis for analysis and criticism. First, we should consider Brazil's recent historical experience to critically evaluate these theoretical assumptions and then turn to the particular policies proposed and implemented by the regime and evaluate their likely impact on economic development, class inequalities, and social welfare.

Da Silva, in terms of both the neoliberal philosophy that guides his economic team and in actual economic practices represents a continuity, extension, and deepening of the disastrous neoliberal policies pursued by the Cardoso regime. On all major political economic issues, debt payments, free markets, privatization, monetarism, da Silva's regime is following the Cardoso regime's failed policies (*Financial Times*, December 20, 2002, p. 2). These policies led to eight years of economic stagnation, profound social inequalities, increased indebtedness, and a near-collapse of a financial system, dependent almost entirely on volatile external flows of speculative capital. If anything, Da Silva's economic policy has extended the liberal agenda by reducing pensions for wage and salaried workers, increasing the budget shares allotted to debt payments, greatly exceeding Cardoso in terms of cuts in the social budget. If Cardoso is an orthodox neoliberal, Da Silva's regime is a Taliban neoliberal.

Da Silva and his Finance Minister Palocci rejected any protectionism, moved to extend privatization, and refused to redress the worst abuses of privatized enterprises. Palocci defends international regulations (WTO policies) as a means to attract foreign investment, rejects protectionism for local industries, and privileges foreign capital in competing for public tenders (state contracts). Palocci argues, 'Brazil doesn't want to close itself. We want to sail the open seas of the global market' (*Financial Times*, January 16, 2003, p. 2). He rejects any state intervention as 'artificial mechanisms' of public financing to stimulate consumer demand among millions of impoverished Brazilians. 'Generating the right conditions, market forces will increase income and corporate productivity,' according to Da Silva's economic czar. This Taliban neoliberal conveniently forgets that it was precisely the 'market forces' in Brazil that created the mass poverty and the worst inequalities in the world over the last 100 years of capitalist expansion.

With the unquestioned backing of President da Silva and the rest of the economic team, Palocci announced the privatization of four state banks, the 'privatization' (*Celso Furtado*) of the Central Bank (under the pretext of 'autonomy' from elected officials) and the promotion of a law that guarantees foreign capital 100 percent control of a substantial sector of Brazil's telecommunication industry, the latter ministry in the hands of the rightwing PMDB as of January 2004. Faced with the failure of AES, the US power company, to meet payments on its purchase of Electropaulo, a power distributor in São Paulo City, Da Silva's economic ministers refused to renationalize the company despite its blatant financial mismanagement (*Financial Times*, February 26, 2003, p. 15).

Dogmatic belief in the virtues of foreign capital as the engine of growth blinds the da Silva regime to the precariousness and vulnerability of tying Brazil's development to international financial capital, as the Brazilian crisis of the late 1990s demonstrated. Domestic austerity and other neoliberal pronouncements were not enough to attract new long-term investment in 2003. By adopting the neoliberal agenda and financial dependence, Brazil will follow one austerity policy after another, austerity without end. The outlook for 2004 is for *further* budget constraints to attract foreign investors.

Economic performance

The economic performance of the Da Silva regime's orthodox neoliberal model was one of the worst in modern Brazilian history and among the worst in all of Latin America for the year 2003. Brazil grew by 0.6 per cent, according to a report of the United Nations Economic Commission on Latin America. (*Argenpress*, December 17, 2003.) Taking account of population growth Brazil experienced a *negative* growth rate of –1 per cent, far below the regime's ideologically informed projections of a 2.8 per cent growth rate, and the second lowest in Latin America.

Unemployment reached record levels in the Greater São Paulo industrial region—exceeding one-fifth of the economically active population: 20.6 per cent in September 2003 (*Folha de São Paulo*, October 24, 2003 B-1). Between January and October 2003 the national unemployment rate grew from 11.3 per cent to 12.9 per cent (*Folha de São Paulo*, October 23, 2003, B4). Among young people (16–24 years) unemployment touched 50 per cent. Moreover, among the new employment positions (772,000) over 92 per cent (716,000) were in the informal sector, lacking social benefits, health insurance, vacations, and job security. Moreover, the average income of a worker in the informal sector is almost one-third less than that of a worker in the formal sector, averaging about $182 a month, far below the poverty line for a family of four. Income levels also dropped precipitously—15 per cent between January and December 2003.

The domestic recession, however, was instrumental in improving Brazil's external accounts with a surplus of $3.856 billion. The surplus was a result of the domestic recession, which significantly reduced imports of consumer and capital goods, and the economic incentives which the regime gave to the agromineral export elites.

Da Silva's orthodox IMF economic strategy premised on an alliance with foreign overseas financiers was not only a complete failure in reactivating the economy, it drove the country further into recession. In contrast, the Argentine regime of Nestor Kirchner, who adopted a heterodox economic policy of limiting the budget surplus to 3 per cent to pay foreign creditors (50 per cent less than Da Silva), created two million subsistence public sector jobs and sustained a *de facto* moratorium on parts of the foreign debt payments in defiance of IMF demands to increase the budget surplus to pay foreign creditors. Argentina insisted on reducing the private foreign debt by 75 per cent and to pay with long-term bonds. Kirchner's economic strategy is to channel public investment toward the domestic market and to promote the small and medium-sized national enterprises as well as the traditional agro-export elites. As a result Argentina's GNP grew by 7.3 per cent in 2003, its industries grew over 10 per cent, and its unemployment declined to 17.5 per cent (counting those on social security payments) down from 22 per cent at the beginning of the year. The point of this comparison is to demonstrate that even within a modified neoliberal framework there are alternatives to Brazil's slavish pursuit of the IMF agenda and the alliance with overseas finance capital—an alternative that reduces debt payments, increases employment, and promotes industrial growth instead of bankruptcies.

In his election campaign Da Silva promised to create ten million new jobs. Instead, after his first year in office, there were an extra one million unemployed workers (*Outro Brazil*, Benjamin et al., November 3, 2003).

The Brazilian state's budget surplus of 4.25 per cent results in over $23 billion in tax revenues being transferred mostly from wage and salary workers to wealthy domestic and foreign creditors (mostly bankers), who in turn invest in speculative activities, principally high interest-bearing bonds from the Central Bank *(Outro Brazil,* Benjamin, November 2003, p. 7). In effect, the Da Silva government policy not only deepens Brazil's already notorious socioeconomic inequalities, but encourages the speculative over the productive market.

Following the lead of their international financial partners and allies the Da Silva regime has implemented a series of regressive 'reforms.' These include pension, tax, and labor legislation designed to increase profits, concentrate capital, and reduce wage and social benefits, all with the hopes of increasing exports and attracting overseas capital. In the tax realm the Da Silva regime has cut corporate taxes and provided long-term tax breaks for foreign investors and tax stimulus packages for agro-exporters while increasing taxes by 27 per cent for wage earners, salaried employees, and pensioners. In effect, Lula's tax policies have 'redistributed' the tax burden from capital to labor, thus further increasing Brazil's perverse inequalities. This is particularly evident in the countryside where agro-export revenues (particularly in soy bean, beef, and citrus) have increased, while the minimum wage for agricultural workers has declined in real terms, in part because of the huge surplus of landless workers. Nothing tells us more about the class character of the Da Silva regime than its fiscal policies—its strident promotion of the export elites and its regressive tax and incomes policy.

The regime's other major triumph in its pursuit of a pure, unadulterated, neoliberal model is found in its public pension reduction and privatization policy. Two aspects stand out: the form in which the Da Silva administration aggressively pursued its policy; and the radically regressive substance of the policy. The regime's savage attack on public pensions was extraordinarily aggressive, demagogic, and highly organized from the top down. The regime extrapolated the 5 per cent highest paid pensioners and manipulated these 'facts' into a general attack on the 95 per cent of pensioners receiving a decent to modest pension. Equally significant, the Da Silva team put pension reduction on the top of its legislative agenda, for both substantive and symbolic significance. Major reductions in pensions augment regime coffers temporarily and provide funds to meet full and prompt payments to creditors even under a stagnant economy. Secondly, pension reductions served to dispel any doubts among overseas speculators about the rightwing character of the Da Silva regime, thus consolidating the strategic ties between the regime and Wall Street. The fact that Da Silva was willing to savage one of his principal long-term organizational and electoral bases of support (the public employees and municipal unions and their millions of supporters) in pursuit of IMF and foreign capital linkages was convincing evidence of Lula's commitments. The final proof of the regime's embrace of big capital was the way in which it disciplined and enforced conformity of its parliamentary representatives. With the exception of three Congress members and one senator, the so-called Workers Party (including its self-styled 'left wing') voted in favor of the regressive pension policies, aided and abetted by the ex-trade union bureaucrats of the CUT. To further demonstrate his strategic ties to big capital, Lula and the PT leadership in Congress expelled the four dissident Congress members and threatened the same to a shrinking minority of 'leftist' Congress members. Da Silva's pension legislation is regressive, not reformist, because it substantially reduces the net payments of those earning over $409 per month for state workers and $492 for federal workers by 11 per cent—the new taxes they will have to pay. A retired federal pensioner receiving $500 a month will now receive $445 a month. For those pensioners receiving over $815 there will be a 30 per cent reduction in income, costing pensioners over $17 billion over 20 years.

The so-called 'pension reform' includes the first steps toward privatizing the multi-billion dollar state pension funds through private investment fund managers. Like Da Silva's so-called 'tax reform,' the regime promises to add a regressive value added tax (in place of a tax on industrial production) and increase tax exemptions for the export elites, while phasing out the mildly progressive financial transaction tax (*Financial Times*, September 2, 2003). The deputy leader of the PT in Congress, Paulo Bernardo, announced that the 'tax reform' would consider guaranteeing tax concessions given to companies by state governments for up to ten years (*Financial Times*, September 2, 2003, p. 2), as the big industrialists and foreign MNCs demand.

In addition, the regime lifted price controls on over 200 basic food and pharmaceutical items, thus increasing corporate profits and reducing living standards for wage and salaried workers. The regime's attempt to forge an 'alliance' with foreign capital fell far short of expectations. The 40 per cent decline in new foreign capital investments suggests that the 'alliance' has not delivered according to the ideological expectations. More seriously, the fundamental economic assumptions underlying the Lula–MNC strategy have been demonstrated to be false. Between 1995 and 2001 foreign investment more than tripled from R272.6 billion to R914 billion. In the same period, unemployment grew 155.5 per cent. In other words, there is an inverse relation between the inflows of foreign capital and employment. There are several hypotheses which could explain this relationship: much of the capital flows were directed at purchasing public or private Brazilian firms, frequently resulting in substantial lay-offs before or during the buyout to enhance profit margins. These privatizations did not necessarily increase production so much as capture monopoly markets (in communications, light and power, and other public utilities). Secondly, much of the new flows of external capital were directed at the paper economy, speculative activity that sought to profit from high interest rates. The increase in speculative foreign investment in high-interest government bonds was accompanied by the bankruptcy of productive firms and decline in productive investments due to the high cost of borrowing which lowered profits in productive sectors. Moreover, many investors switched from investments in risky and stagnant productive sectors to the lucrative speculative sector—all of which contributed to increasing unemployment. Finally, the 'free-market policies' led to a big increase in cheap and subsidized imports displacing local small and medium-sized agricultural and manufacturing producers who employ the bulk of the labor force. The growth of imports in these labor-intensive sectors led to the growth of a large 'surplus labor force' in both major urban cities and in the countryside.

The pattern of large inflows of foreign capital and increasing unemployment under Cardoso intensified under the Da Silva regime, with its exaggerated dependency on foreign investors. The regime's record transfer of Brazilian earnings overseas (profits, interest and royalty payments, service charges, domestic legal and illegal transfers to foreign accounts) surpasses $50 billion—enough to fund a major job-creation program of public investment, an agrarian reform to settle 200,000 landless families, a comprehensive health program, doubling of the education budget, and a real 'zero hunger' program which would cover the tens of millions of impoverished Brazilians who have yet to be reached by the current failed program.

The current reduction in public investment and purchasing, the sharp drop in disposable income among wage and salaried workers, the decline in agrarian reform beneficiaries, and the troubled state of most cooperatives for lack of credit mean that Brazil's domestic market is an increasingly unattractive sector to invest in, except at the high end of luxury goods and in the agro-mineral export enclaves. Da Silva's class-based austerity program designed to attract foreign capital is likely to limit new investments to selected export sectors in agro-business and industry, speculative, financial, and banking activities, luxury production, and commercial activity (import houses).

The official projections of 2004 are for 'modest' growth of 3.5 per cent; but this is a dubious proposition for several reasons. It assumes large-scale foreign investment based on low inflation, budget surpluses, and tight monetary policies. But given the Da Silva regime's long-term commitment to a 4.5 per cent budget surplus, internal demand will continue to stagnate. The regressive tax, income, and labor policies will continue to weaken mass demand for consumer goods. In effect, the projection is for Brazil to grow in the capital-intensive, low-labor export sector and regress in the rest of the economy, deepening socioeconomic inequalities and widening economic disparities within and between sectors of the economy. Equally important, the Brazilian economy floats in stagnation in large part because of exceptionally favorable commodity prices (especially in iron, soy beans, etc.) that are very vulnerable to great fluctuations. A sharp drop would diminish the 'trade surplus' and affect the availability to finance the agreed to exorbitant debt repayments, leading to the flight of capital or at least a sharp reduction in capital inflows. The result would be a profound recession and perhaps collapse of the regime's financial architecture.

President Da Silva, in one of his most arrogant and self-serving declarations, called his presidential predecessors 'cowards' for not ignoring the basic health, welfare, and employment needs of millions of working Brazilians. In his perverse logic, supporting the rich and powerful (especially the foreign rich) backed by the major media monopolies was an act of 'courage.' Lowering the minimum wage of the poor and destitute, reducing the pensions of the public employees, and weakening protective labor regulations, abandoning 200,000 encamped landless rural workers, and then telling the Brazilian public that he speaks for the working people require great audacity—the courage sincerely to articulate and repeat the Big Lie.

Brazil's export growth partially accounted for the $25 billion trade surplus. Equally important was the stagnation of imports, which grew by only 2 per cent in 2003, slightly higher than population growth. Agro-mineral products ($33 billion) accounted for almost half of Brazil's $73 billion exports. Brazil's exports were primarily directed toward the European Union ($18 billion, or 25 per cent of the total) followed by the US ($16.9 billion or 23 per cent) and China ($4.5 billion, or 6 per cent). Its principal trading partner in Mercosur, Argentina, accounted for $4.5 billion, or 6 per cent. In other words, the so-called Latin American regional organization accounted for less than 10 per cent of Brazil's trade, hardly an example of a 'regional bloc' to challenge ALCA or the World Trade Organization. Clearly, Mercosur, despite Da Silva's rhetoric, is a very subordinate part of his trade strategy, which is oriented to complementing the European Union and integrating ALCA—if the US plays the 'free market' game.

Da Silva's policy toward consumers and health issues follows directly from neoliberal dogma and in complete opposition to the expectations of his popular followers.

The regime approved new price-hikes by privately owned utilities, thus increasing the burden on the poor (*Financial Times*, February 18, 2003, p. 4). In February 2003, Da Silva eliminated price controls on 260 pharmaceutical products and proceeded to liberate 3,000 medicines from price controls in June 2003.

In a bizarre twist, to compensate for declining living standards, Lula promised to install 4,200 computers for the poor and give them ten minutes free time daily. Given the price–wage squeeze on wage earners and the potential for discontent, Lula is ensuring the loyalty of the police; he granted them a 10 per cent salary increase.

It is no wonder that Da Silva received thunderous applause from the super-rich in Davos in January 2003. As Caio Koch Weser, Germany's State Secretary of Finance, told him: 'The key is that the reform [neoliberal] momentum gets the benefit of the enormous credibility that the president brings' (*Financial Times*, January 27, 2003, p. 2).

Da Silva's deliberate manipulation of his working-class origins to promote a big business agenda was and is much appreciated by the shrewd financiers on both sides of the Atlantic.

ALCA and US imperialism

Throughout Latin America mass popular movements are vociferously protesting against ALCA. Millions of peasants in Mexico, Ecuador, Colombia, Bolivia, Paraguay, and Brazil have blockaded highways and demanded that their governments reject ALCA. In Brazil in 2002 a referendum was held on ALCA in which over ten million participated and over 95 per cent voted against. Da Silva refused to participate and ordered the PT not to become involved. Once elected, he ignored the ten million voters against ALCA and agreed to be a co-partner with the US in the negotiations to consummate agreement on ALCA.

ALCA is a radical comprehensive trade agreement which, if implemented, would transfer all trade, investment, and other economic policies to a US-dominated economic commission, probably located in the US, which would oversee the privatization and US takeover of the remaining lucrative state-owned public utilities, petroleum, gas, and other strategic industries. At a speech at the National Press Club in Washington, Da Silva pledged to create a western hemisphere trade pact. He promised to push ahead with ALCA and was ecstatic about his relation with President Bush: 'My impression of Bush was the best possible' (*Financial Times*, December 11, 2003, p. 5).

Da Silva and his economic team's main objection to ALCA is that it must reduce trade barriers for Brazil's big agro-exporters. The 'worker president's' embrace of the most aggressive militaristic US president, at war with Iraq and plotting the overthrow of the democratically elected government of Venezuelan President Hugo Chavez, must indeed touch the lowest point of political servility in recent Brazilian diplomatic history.

As many critical economists have demonstrated, ALCA will destroy family farmers and peasant agriculture, increase the number of landless peasants, spreading hunger and forcing mass migration to the urban slums, making a mockery of Da Silva's 'zero hunger' program. Lula's pitiful handouts of temporary food relief will not compensate for the millions of new poor and destitute resulting from his doctrinaire neoliberal policies. Da Silva claimed that his 'zero hunger' scheme was 'much more than an emergency donation of food. We need to attack the causes of hunger, to give fish and to teach how to fish' (*Financial Times*, January 31, 2003, p. 2). Instead, with ALCA, da Silva is attacking the poor, not hunger, and strengthening and deepening the causes of hunger, not lessening them.

In pursuit of the best possible relations with President Bush, Brazil's Foreign Minister Celso Amorin sought to intervene in the Venezuelan conflict. Amorin offered to 'mediate' in the dispute between the constitutionalist President Chavez and the US-backed 'Democratic Coordinator,' by organizing groups of nations dubbed 'Friends of Venezuela.' The so-called 'Friends' included Spain and the US, both of which supported the April 11, 2002 failed coup against Chavez. In addition, the 'Friends' included the neoliberal regimes of Chile, Mexico, and Portugal and of course Brazil. President Chavez, who belatedly became aware of Amorin's trap, asked that a few more friendly countries be included. Da Silva and Amorin refused and the Brazilian ploy on behalf of the US-backed opposition became a dead letter. Chavez told the 'Friends' and their Brazilian sponsors to keep out of Venezuela's internal affairs. This did not prevent Amorin from declaring that the Brazilian regime was open to meet with the Venezuelan *golpistas* (*La Jornada*, January 22, 2003).

The key to the Da Silva regime's policy toward ALCA can be found in discussing the principal protagonists of this economic strategy: the IMF, foreign capital, and the agro-mineral and big industrial sectors. Lula's economic team (Mereilles, Paolucci, Furlan, Rodriguez, and their academic advisors and financial backers) have rigidly pursued a coherent economic strategy that is based on attracting large flows of foreign capital, promoting agro-mineral exports, and gaining access to international capital markets by over-fulfilling its surplus to meet public and foreign debt obligations. This is an export-driven strategy *par excellence*, but one that differs substantially from the Asian model in several important aspects. The regime does not pursue an 'industrial policy' of selecting and financing national public and private manufacturing capital—public productive investment declined. Secondly, the Asian countries did not channel anywhere near the percentage of its GNP to overseas bankers—they invested far more of the economic surplus in domestic production. Thirdly, the Asian export growth strategy protected 'non-competitive sectors' from subsidized imports in contrast to Brazil, which is willing to negotiate the opening of its domestic market in exchange for access for its exports. Fourthly, the Asian export strategy was based on the export of manufactured goods, with a high content of value added, while Brazilian leading exports, mostly agro-mineral, have little value added, thus not increasing domestic employment and expanding the domestic economy. Fifthly, the lead export enterprises in Asia were national-private or publicly regulated, whereas in Brazil, most are privatized and many are foreign-owned, leading to lower rates of re-investment in the Brazilian economy.

As a consequence Da Silva's 'export strategy' has a vastly different outcome from that which occurred among the Asian Tigers. Export growth sectors are confined to enclaves with few spread effects; growth is uneven and unstable, dependent on highly volatile markets and commodities. Income and employment do not change significantly for the better. In fact, it is worsening given the new 'flexible' labor policies. A key difference between the Asian export model and Da Silva's economic team's is in the timing: the Asian export model began with a strong national industrial protectionist model and later moved toward liberalization, while Da Silva starts with an orthodox liberal approach which undermines any efforts to build up competitiveness. The selective liberalization in Asia allowed the regimes to regulate the inflows of capital toward productive activity, while in Brazil the indiscriminant liberalization has encouraged the entry of mostly speculative, not productive capital—in part because of the higher rates of return in the former over the latter.

During the Cardoso years there was a tidal wave of foreign investment, a substantial part buying public and private national enterprises. The Census of Foreign Capital taken in 2001 revealed that Brazilian enterprises having at least 20 per cent foreign capital increased from 6,322 firms to 11,404—an increase of 80.4 per cent—between 1995 and 2000. The stock value of firms with foreign capital increased three-fold in the same five-year period. The liberalization policy that encouraged foreign inflows did not decrease unemployment; in fact, unemployment increased 15.5 per cent between 1995 and 2000. The Da Silva regime has expanded the concession to foreign investors but has reaped meager results as new investments have declined, pending the privatization of the remaining lucrative public enterprises. Foreign direct investment declined by half in 2003, thus weakening one of the major protagonists in the Da Silva 'export model.' Several reasons account for this.

First of all, many of the most lucrative public enterprises which attracted foreign capital between 1995 and 2001 have already been sold; few enterprises are left. Secondly, the rates of return in the speculative, financial, and stock markets are several times greater than the rates of return in the productive sector. Thirdly, the stagnation of the Brazilian economy, the high interest rates, and declining purchasing power of most Brazilians is not an 'attractive' market to invest in; the labor costs in Brazil, while among the lowest on the continent, still are higher than those in China, for those manufacturers looking at export sites for industrial investments. Despite being considered 'the toast of financial markets' (*Financial Times*, December 1, 2003, p. 14), Da Silva's single-minded pursuit of the export surplus has lowered earnings 15.2 per cent while bankruptcies rose throughout the year and foreign investment fell. The next stage in 2004 will be to deepen and extend the regressive policies ('reforms') by decreasing regulations over energy and telecommunications (increasing prices to consumers), passing more labor 'reforms,' lowering severance pay, increasing precarious employment, easing firings to 'attract' foreign investors. In other words, the orthodox neoliberal measures taken in 2003 will be radicalized, as the regime will dogmatically pursue the illusion of great waves of investors, reactivating the economy.

The second major component of Da Silva's export strategy is to comply with the IMF and World Bank in hopes of securing 'certification' among overseas lenders. The cost of Da Silva's pact with the IMF is enormously costly to the economy and a scourge to any development, dependent or not. The Da Silva economic team has agreed to pay the IMF over $35 billion during the four-year term of his presidency. Rather than attracting new capital, the financial agreements with the IMF, particularly the setting of a budget surplus of 4.25 per cent of the GNP to be paid to foreign creditors, means that over $27 billion left the country—nearly three times the size of the inflows of foreign investment. This counter-growth strategy adopted by Lula's team is projected to continue through to the end of his term by Finance Minister Palocci (*Folha do S. Paola*, October 31, 2003, B5). The prospects for 2004 are hardly auspicious as an increase in imports is likely to shrink the trade surplus, and foreign debt amortizations will increase from $27.4 billion in 2003 to $46.9 billion in 2004 (*Financial Times*, November 7, 2003, p. 3). Da Silva's orthodox economic policies have entered into a vicious circle: the more the regime borrows, the harsher the conditions, the weaker the growth, the lower the investment, the greater the debt obligations to GNP. What Da Silva's economic team refuses to recognize is that foreign investment does not *create* expanding markets or industrial growth; rather, foreign investment is *attracted* to expanding markets and industrial expanding economies.

The third aspect of Da Silva's export growth strategy is the search for markets through ALCA and new trade partners. Da Silva traveled to over 27 countries mainly as a salesman for the agro-mineral and industrial elites. Given the centrality of the 'export-led' strategy of the Da Silva regime, and its close structural links with the giant export enterprises in agriculture, mining, and petroleum, it is not surprising that the regime has been a staunch supporter of comprehensive liberalization of trade. Contrary to most 'leftist' opinion, Brazil's 'leadership' of the G-21 at the Cancun meeting had nothing to do with the defense of the poor and downtrodden of the Third World. The main bone of contention was Brazil's militant defense of its agro-export elite's open access to US markets. Da Silva has reiterated time and again his favorable position on 'free trade' as the road to growth and prosperity (despite the devastating impact on Brazil and the rest of Latin America over the past two decades). Celso Amorin, Brazil's neoliberal Foreign Minister, insisted that the US eliminates its tariff barriers, quotas, and subsidies that hinder Brazilian exports of sugar, cotton, soy bean, beef, and citrus products. The US protectionist position defended by US Trade Representative Robert Zoellick was unacceptable to Brazil because it put in question the entire Da Silva free market export-led strategy. The ALCA meeting in Miami of November 17–21, 2003 led to a compromise in which the Brazilian Foreign Minister agreed to set aside Brazil's objections to US protectionism and agricultural subsidies in exchange for an agreement that allows member countries to opt out of parts of the agreement they found objectionable. The US thus was able to push its agenda creating a legal-political framework for opening up competition on government procurement, defense of intellectual property rights, liberalization of services, lowering of subsidies and protection (in Latin America), and 'equal' treatment between giant US MNCs and smaller Latin American firms. What Celso Amorin called 'ALCA Lite' is in fact a substantial step toward consummating the US version of ALCA—and its *de facto* colonization of Latin America. US imperial strategy operates on two levels: the signing of regional and bilateral 'free trade agreements' with its Andean regime clients (Ecuador, Colombia, Peru, Bolivia) which readily accepted its one-sided liberalization, the Central American clients (Honduras,

Nicaragua, El Salvador, Guatemala) plus bilateral agreement with the Dominican Republic, Chile, and Mexico. The US strategy is to use the pressures of these free trade agreements to push Brazil and Argentina into signing up to ALCA in order to retain markets in the neighboring countries.

The dilemma of the Da Silva regime is that while counting on the US market and investments it is strongly linked to its own agro-exporters. These contradictory pressures find expression in the efforts of Amorin to accept ALCA in stages, hoping to lower some of the trade barriers in exchange for conceding to the US demands on services, investment, and intellectual property rights (all of which have very damaging impacts on Brazilian industries, including the pharmaceutical, finance, health and insurance sectors). Apart from ALCA, Da Silva has traveled throughout Asia, Africa, Europe, and North America, in active pursuit of markets for Brazilian exporters, usually accompanied by an entourage of big business people. In contrast to the tireless attention Da Silva has given to the interests of the agro-mining elites, he ignores the hundreds of thousands of landless rural workers encamped along the highways under plastic tents, telling them to wait, to have patience, to suffer nobly.

Brazil's promotion of Mercosur is also part of an effort to diversify and extend markets—a greater necessity now particularly with declining domestic consumption and agricultural constraints in Europe and the US. But in the case of Mercosur Brazil competes with Argentina in agricultural, textiles, and beef products and runs into a depressed market. The big push of the Da Silva regime is toward Asia, particularly China, where iron and soy bean exports are experiencing double-digit growth. China's boom economy, particularly its manufacturing and consumer sectors, have pushed the price of iron and soy beans to exceptional levels. As a result, the regime is encouraging big agro-export growers to expand their areas of cultivation, even to precarious regions—like the Amazon—and has turned a blind eye to the displacement of Indians and small producers.

Export industries are the 'nobility' in the Da Silva conception of development, while local producers are the 'vassals.' The urban poor, landless rural workers, and the unemployed workers are the 'serfs,' to provide cheap labor and services, consume less and keep quiet, in order to service the divine justice of increasing export earnings and fulfilling foreign debt payments. Together, foreign investors, exporters, and financiers form the Unholy Trinity guided by the courageous helmsman Inacio Da Silva, who bravely tramples on the poor and worships at the feet of the IMF.

This is why the financial journals proclaim the 'success' of Da Silva's export development strategy. The billions spent in promoting exports are based on keeping the minimum wage of the Brazilian worker at $87 per month. Read the export figures and ignore the growing lines of unemployed (150,000 in Rio applied for 1,000 street cleaning positions) and the misery of millions of landless rural workers. As one MST militant told me: 'Exports are doing great, only the people are suffering.'

An investor-speculator paradise

By late 2003 the big news on the financial pages of the Brazilian press was the fall of productive investment and the growth of speculative capital—especially foreign capital (*Jornal do Brazil*, October 24, 2003, A22). Most of the foreign investment went into speculating on the Brazilian stock market, and purchasing government bonds with one of the highest rates of interest on the planet (18.5 per cent as of December 2003).

The boom in highly volatile speculative capital is based on short-term considerations—high interest rates and an overvalued currency, which is strangling the domestic economy and precluding the revival of the economy. Investor banks reported a 'boom in emerging market debt' (*Financial Times*, December 24, 2003, p. 13). With relatively stable neoliberal regimes in power in the major countries of the Third World subordinating all priorities to meeting their debt obligations, there was record investment in debt. Speculators invested in a record $3.3 billion in Third World bonds, doubling the figure for 2002 ($1.7 billion). The best performer from the point of view of the speculators was Latin America, where profits reached 35 per cent, while in Asia, engaged in a more independent productive trajectory, the rate of return was 12 per cent (*Financial Times*, December 24, 2003, p. 13). Within Latin America, Brazil was the most lucrative country, generating one of the highest returns in the world.

The year 2003 was a period of spectacular profits and gains by bond and stock market speculators and overseas investment houses. The regime's economic and fiscal policies were tailor-made to benefit the most parasitic sectors of the economy, those commodity sectors geared to overseas markets and the most rapacious overseas speculators. While bankruptcies soared and the producers for the domestic economy fell into a profound recession, the Brazilian main stock index rose to its highest level since it was founded 35 years ago in 1968. Da Silva's regressive social policies, tight monetary policies, promotion of agro-export elites, budget and trade surpluses provided ample incentives for the boom in the stock market. The stock market index (BOVESPA) rose from 11,268 in January 2003 to over 20,000 by the end of the year—one of the highest increases in the world (*Financial Times*, November 28, 2003, p. 19). The same is true with regard to returns on bonds. Brazil's market returns rose 60 per cent during 2003—three times the composite index for all 'emerging market countries.' Stock market speculators were able to reap an 80 per cent rate of return; bondholders saw their profits increase by over 60 per cent under the most favorable conditions of any country in Latin America. The reasons for this 'speculator's paradise' is the direct result of the regime's economic policies: the harsh budget cuts, the reduction of salaries and public employees' pensions, the huge 'budget surplus' of 4.3 per cent—these were directly or indirectly transferred to financiers, speculators, and bondholders. Overseas investors are the point of reference for all the major economic decisions taken by Lula's Finance Minister Palocci and the Central Bank as they have readily admitted. This 'carnal relationship' with the financial speculator sector is not a temporary or conjunctural phenomenon. Palocci announced on December 13, 2003 that a budget surplus of 4.25 per cent would continue to be the regime's policy for the next ten years—an optimistic assessment of the 'Workers Party's electoral future.

In pursuit of this alliance with local and overseas financial and speculative interests, the Da Silva regime is pushing hard to secure legislation making the Central Bank 'autonomous' from the legislature—thus increasing its ties to the big financial groups. The ties with finance and speculative capital are justified as 'gaining investor confidence' and 'securing economic stability' in order to stimulate future growth. The fiscal deficit as a percentage of GNP actually rose from 4.7 per cent in 2002 to 5.3 per cent in 2003, due to exorbitant interest rates on the internal public debt and negative economic growth.

The massive inflow of speculative capital in the paper economy was matched by the decline of private investment, including foreign investment, in the productive sectors and the sharp decline in public investment. Given the high interest rates, shrinking domestic market caused by 21 per cent unemployment in greater São Paulo, it was less risky and more lucrative to speculate on the stock market and government paper than to invest in productive sectors, particularly the manufacturing sector. In effect, the Da Silva regime's policies have led to the de-capitalization of the productive sector and the over-capitalization of the speculative sector—a formula that is clearly not sustainable in the medium term. Instead of following a policy of strengthening the national industrial bourgeois, Lula's policies are converting productive capital into speculative, thus reinforcing the rule of the overseas and local financial oligarchy. If the regime is a 'government in dispute' as some of the 'leftists' in the PT claim, it is not between capital and labor; it is between speculative and industrial capital.

Having anchored state policy to the volatile behavior of financial and speculative capital, the regime has sharply reduced its options for the future. A sharp break would disrupt financial activities; a continuation of pro-speculative policies will perpetuate stagnation and heighten the chances of a major economic collapse.

Given the powerful structural linkages between finance capital and the regime and in line with its wholly negative policies toward labor, the peasantry, public employees, and the urban poor, the regime is clearly a government of the *right*. A rigorous comparison with parties of the 'center-left' (social-liberal), or of the center-right, demonstrates that the regime lacks any program to increase welfare legislation, to develop a national industrial policy, or even to promote greater consumer spending.

Even on the basis of the regime's own goals of attracting long-term foreign investment in productive sectors, they failed: despite all the concessions and the servile implementation of the IMF formulas, direct foreign investment fell to less than one-third of its level during the Cardoso years. On the other hand, the Da Silva regime has followed a coherent and radical neoliberal agenda that is completely in line with the policies of the most retrograde sectors of finance capital. The fact that speculators were able to increase their earnings by 113.6 per cent betting on the stock market in 2003 is emblematic of the Da Silva regime's real political identity (*Financial Times*, November 28, 2003, p. 1).

The boom in speculator investment in the Brazilian stock market has little to do with Da Silva's long-term policy and is more the result of excess liquidity in international capital markets. In other words, an 'external shock,' such as has frequently occurred in the recent past, could put an end to the stock market bubble.

Despite the massive debt payments, Brazil's total public debt increased from R893.3 billion in 2003, to R965.8 billion in 2003—about 8 per cent. For 2004, public debt amortization will total $37 billion. (*Financial Times*, January 16, 2004). Given the stagnant economy and the compression of living standards, many economists question the long-term sustainability of Brazil's public debt. While the Central Bank reduced the inter-bank interest rate to 17.5 per cent in November 2003 (from a high of 26.5 per cent in May), the market interest rates are prohibitively high. The average cost of borrowing is 71.3 per cent while the average for personal loans is 149.3 per cent (*Financial Times*, November 21, 2003, p. 3). These rates are not likely to stimulate any muscular recovery in 2004.

Agro-exports and agrarian reform

Agrarian policies illustrate the priorities and class nature of the Da Silva regime more than any other sector of the economy. The agro-export sector controlled by a tiny elite of plantation growers and agribusiness MNCs experienced spectacular growth thanks to lucrative subsidies and tax incentives. In contrast the landless rural workers, the cooperatives and family farmers suffered the worst year in recent memory in terms of land distribution, rural credits, and technical aid.

According to a year-end evaluation by the Rural Landless Workers Movement (MST):

> During this year, the government did very little for Agrarian Reform. There were hardly any expropriations. Government credits via PRONAF were few and above all their form of application never reached the land reform settlements, which spent the year [2003] practically without resources. There were few projects in the land settlements. Few states contracted technical assistance for the land settlements. There was a great deal of bureaucracy and incompetence in INCRA, the land reform agency. (MST, *Biblioteca de Artigos Tematicos*, December 2003)

Lula had promised to settle 60,000 families on expropriated land in 2003 but finished the year with 10,000 families. The MST had demanded the settlement of 120,000 families; Da Silva met 12 per cent of the MST target. To put the immense failure of the regime in perspective, it is useful to compare the figures to the previous regime, which averaged 40,000 families per year over an eight-year period. In other words, the Da Silva regime barely reached 25 per cent of the previous regime's dismal annual record, a great leap backward for the agrarian reform movement, at least from the point of view of the landless workers. Moreover, if we subtract the number of land settlers who were forcibly evicted by the state judicial authorities (9,243 families), the net land reform beneficiaries is less than 1,000 in a country of 4.5–5 million landless families, numbering between 25 and 30 million impoverished rural people.

Over 200,000 families living in the most precarious conditions by roadsides and abandoned fields have dismal prospects for the immediate future unless they take the initiative and organize land occupations. The basic reasons for the failure of the regime to implement land reform is because of the priority he has given to paying the foreign debt, meeting the austerity targets of the IMF, and to promoting the agro-export sector. Funding for human rights issues including land reform have the lowest priority.

In October 2003, having clearly failed to meet his promises to the landless workers and having openly sided with the big agro-exporters, Da Silva embarked on a scurrilous and unprincipled attack on the MST and its agrarian reform proposal. 'I am not going to carry out the agrarian reform which the MST is proposing, trading urban misery for rural poverty, simply to boost the number of land reform beneficiaries who produce nothing' (*Veja*, October 29, 2003, p. 40).

Contrary to Lula's bombast, over the past 19 years 350,000 land reform beneficiaries not only produced millions of *reales* worth of food each year for the local market, but also have developed export products. Moreover, almost all objective scholars and journalists have noted the vast improvement in the lives of the land reform beneficiaries. In fact, Da Silva has no alternative land reform, as the ex-president of INCRA, Macelo Rezende, specified when he announced his resignation in August 2003.

The only agrarian reform which took place in Brazil resulted from mass direct action from below. Land occupations increased from 176 in 2002 to 328 in 2003, an increase of 86 per cent. The number of organized campsites of landless workers preparing for land occupations increased from 64 in 2002 to 198 in 2003, an increase of 209 per cent (CPT Document, December 21, 2003). The number of families that participated in land occupations rose from 26,958 in 2002 to 54,368 in 2003, an increase of 102 per cent. The number of families organized in campsites in 2003 rose to 44,087 against 10,750 in 2002, a 310 per cent increase. Between January and November

2003 there were 1,197 rural conflicts compared to 879 in 2002, an increase of 36 per cent.

The landless workers no longer believe Lula's promises; they are taking matters into their own hands and moving forward. The total absence of government support and initiatives has led to increased class conflict and an increasing reliance on direct action. On the other hand, Lula's positive support for big agro-export farmers and their allies among the judiciary has led to scores of arrests of rural activists and the assassination of dozens of activists (at least 80 by December 2003).

During the first year of the Da Silva regime, agrarian reform has been dismal by any measure. UN human rights observers, Brazilian Church (CPT), and human rights activists have recorded growing landlord violence and extra-judicial executions, state criminalization of social movements, arbitrary arrests, and the continued impunity of police torturers and assassins. The fundamental explanation lies in the deep continuity of the judicial, police, and administrative apparatus from the past and Lula's refusal to acknowledge the unequal and selective implementation of law. The regime's *de facto* criminalization of the social movements aids and abets the local landlords in their bid to extend their para-police activities.

The second reason for the regime's dismal human rights record is Lula's economic team's deep commitment to creating a 'favorable climate' for foreign investors—and his determination to repress any signs of social protest as a 'threat to social peace.'

The third reason is found in the regime's agro-export strategy. Given the high priority which the regime gives to meeting the demands of external creditors and fulfilling Lula's agreement with the conditions of the IMF, his regime favors those sectors of agriculture that generate hard currency at the expense of those agricultural sectors producing food for local consumption. It is precisely the 'triple alliance' between the regime, the agro-export elites, and external financial creditors that has undermined the regime's commitment to agrarian reform. It is the triple alliance that has led to the regime's commitment to negotiate Brazil's entry into ALCA under conditions that permit entry of Brazilian agricultural exports into the US market in exchange for the free entry of US food imports that bankrupt local producers.

To sustain this 'model,' the regime has opposed demands for agrarian reform and criminalized social movements promoting agrarian reform while pressing the US to lower its tariffs and eliminate its quotas on soy bean, citrus, cotton, sugar, and other export staples.

The problem of human rights violations in Brazil is not simply the result of local officials or landlords but a deep structural problem embedded in the basic strategy of the regime. The elites of Latin America have certainly recognized the value of Lula's strategy. The *Folha de São Paulo* (October 29, 2003) featured on its front page a poll among the elite of six Latin American countries which chose Lula as the 'best President in Latin America'—exceeding all other neoliberal presidents by a wide margin.

Under conditions, as in Brazil, where the entire economic team is composed of officials with structural links to foreign multinationals and domestic agro-export elites who embrace orthodox neoliberal ideology, there is no possibility of 'disputing for power in the regime.' The year 2003 demonstrates that the advocates of the 'inside strategy' failed to secure progressive social changes. Moreover as the year progressed, the orthodox neoliberals moved further to the right, allying with traditional rightwing parties and extending their policies to all spheres of society and economy. Moreover, given the neoliberal regime's centralized control of the PT and parliament it was and remains impossible to expect any social changes from parliamentary or electoral activity. The only positive changes took place via direct action, extra-parliamentary activity, land occupation, strikes, and demonstration.

The major protagonists and principal beneficiaries of Da Silva's financial aid are the agro-business elites. Agricultural exports had a 'boom year,' growing by over 30 per cent according to the Commerce Minister, Luis Furlan (*Financial Times*, July 2, 2003), and owner of one of Brazil's biggest food-processing plants.

Brazil posted a record trade surplus in 2003 of nearly $20 billion, in part because of a boom in commodity prices of soy bean, iron, and other prime materials, a decline in imports because of the drop in living standards, as well as the economic incentives and deregulation introduced by the Da Silva regime. Exports were promoted through generous subsidies that favoured exporters over domestic producers, eliminated regulatory measures on foreign investment, provided 20 'priority industries' (large-scale exporters) with preferential loans at lower, subsidized interest rates, and exempted exports from a host of taxes, thereby shifting the tax burden onto wage workers and producers for the local market. The net effect of the interventionist policies was to increase profits and opportunities for exports, mainly in the agro-mineral sector, while prejudicing the small producers and landless workers. Clearly, the lowering of wages, a low minimum wage and the weakening of the unions decreased the costs of labor and increased profits in the 'dynamic' export sector. The 'export surplus' will not be recycled within the economy to support multi-sector growth, as much of the surplus (in hard currency) will be used to pay foreign and domestic creditors and speculator bondholders. There are few if any 'spread effects' from the export sector to the domestic market. Moreover, the success of the regime's policy of export subsidization is leading to the greater centralization and concentration of capital and land, as well as the expansion of export crops into the Amazon, thus destroying valuable ecological regions. The result of the highly capital-intensive agro-mineral growth is to increase the poverty of the marginalized small farmer and unemployed landless rural workers, and predictably making a sham of Da Silva's 'zero hunger' campaign.

Via Campesina, the most inclusive large-scale, international organization of small farmers and peasant organizations, criticizes the agro-export strategy promoted by Lula:

Giving priority to production for export above production for local and national markets, leads to scarcity of food at the local level and provokes a separation between food, agriculture, fish and its important social dimensions. (Declaration of Via Campesino and Coalition, December 12, 2003)

The documents correctly criticized the so-called G-20 dissident countries led by Brazil, who questioned the western powers at Cancun in November 2003:

Even those governments that questioned the agenda of the European Union and the United States at Cancun continue with the negotiations which basically prioritize export-oriented agriculture.

The document points to the fact that the agro-export strategies of regimes like Da Silva are willing to sacrifice domestic food producers in order to access markets for their staple exports:

In many countries, especially in the South [read Brazil], peasant production is being substituted by low price imports from other countries and by agro-industrial production oriented to exports, using cheap labor and taking advantage of the lax enforcement of social and environmental norms.

The document pointedly attacks the Brazilian leadership of the G-20:

> While the G-20 is a necessary political counterweight to the USA and EU, it principally represents the exporters of the South, and does not defend the interest of the great majority of farmers and peasants who produce for local markets. Moreover [the G-20] has weakened its objections . . . to the US and EU subsidies to their agro-export agriculture and that's why they pursue greater liberalization of the agricultural markets in the South.

The document identifies the conflict over agricultural policy as a class struggle rather than a 'North–South' conflict:

> The true conflict—around food, agriculture, fishing, employment, environment and access to resources—is not between the North and the South but between the rich (agro-exporters) and the poor (the peasants and farmers producing food).

The approach of Via Campesino to hunger is diametrically opposed to Da Silva's agro-export practice and his failed 'zero hunger' program:

> From the broad perspective of national and local economic development it is much more important to confront poverty and hunger by providing resources in a sustainable manner and producing in the first place for the local markets than to export.

Lula's agricultural strategy is logical, coherent, and catastrophic for peasants, landless workers, farmers, the environment, and the native peoples. The agricultural policy is built around a strategic structural alliance with foreign bankers, agrobusiness elites, and multinational trading corporations. The logic is to promote agro-export elites to generate hard currency, to increase trade surpluses that will be used for prompt and full payment to foreign and domestic creditors. This will create confidence in foreign markets and lead to large flows of foreign investment ensuring future growth and rising income. From conception to execution the agro-export strategy is elite and foreign-driven. Poverty and employment is seen as a by-product, a 'trickle-down effect' of supply-side economics. In practice the strategy increases unemployment and sharply decreases real income and the number of land reform beneficiaries. After year 1 poverty and hunger actually rose in Brazil over the previous year, and the compensatory 'zero hunger' program failed to make any comprehensive and sustained impact.

The environment

The first months of Lula's regime revealed the two-faced nature of his environmental policy. He appointed a progressive environmentalist, Marina Silva, as Minister of Environment, and proceeded to slash by 12 per cent the ministry's funding, thus severely limiting its capacity to protect the Amazon (among other locations) from the constant depredations of agro-exporters, principally soy bean, and lumber barons. Lula's environmental record is as bad or worse than that of his predecessors.

Throughout the world, from Western Europe to India, from Africa to Brazil, farmers, peasants, ecologists, and consumers have battled against the big agro-business corporations trying to impose genetically modified seeds and their chemical fertilizer and herbicide packages. Prior to the Da Silva regime, GM-based farming was limited to isolated regions of southeastern Brazil. Without consulting Congress, or the representative organizations of small farmers and landless workers, or environmental groups, the regime decreed the approval of GM seeds, at the request of Monsanto. Despite the opposition of a majority of Brazilians, Lula's economic team, led by its Minister of Agriculture, proceeded to impose the measure. The specter of chemical-based export agriculture threatens to undermine the precarious cost/price margins of small producers and may also prejudice exports to European markets. The spread of high-cost, chemical-based agriculture is leading to the bankruptcy of millions of local producers. Apparently, Lula's commitment to chemically-based agro-export elites overshadows the terrible fate facing peasant farmers.

The second element of the regime's policy of environmental degradation—particularly the Amazon rainforest—is the reduction of personnel, funding, and resources for policing the Amazon region. With a 12 per cent cut in budget, the already inadequate regulatory regime declined in effectiveness, and clear-cutting of the rainforest expanded. Under Da Silva's leadership, Congress voted to reduce the rainforest by 50 per cent of its current size. To promote the expansion of agro-export crops, livestock, and lumber interests, the 2004 budget is giving high priority to expanding road and highway building through the Amazon. Public investments projected for 2004–7 are on the order of R189 billion ($63 billion). To encourage the agro-mineral sector, the regime plans to allocate R58.6 billion to promote exports. The environment is scheduled to receive R6.4 billion (about 10 per cent of the funds for agribusiness expansion). Lula's public investments program clearly favors those economic sectors most involved in exploiting non-renewable resources, most destructive of the rainforest and the Amazon in general, and those sectors likely to displace small farmers and encroach on lands reserved for Indian communities. Lula once more demonstrates his 'courage' in catering to the interest of the most powerful MNCs, the most ecologically destructive industries and the most dubious practices of huge foreign-owned GM corporations at the expense of the poorest of the poor—the impoverished and vulnerable Indian communities and small subsistence farmers in the Amazon (*Adital: 'Plano Plurianual desacredita preocupacao ambiental do Governo brasileiro,'* November 24, 2003).

Lula's policy is accelerating the process of turning a large swathe of the northern Amazon into large scale soy bean plantations, particularly in Para, where they compete with the timber barons in stripping precious, old growth timber like mahogany into highly lucrative, if illegal, exports. It is no surprise that one of the PT's Congress members in Para, Airton Faleiro, is himself a big soy bean grower. The numerous national and international appeals from ecologists, Via Campesino, intellectuals, MST, and Amerindian organizations have fallen on deaf ears. Even worse, the regime resorts to chauvinist rhetoric attacking 'foreigners,' accusing them of trying to impose restrictions on Brazil's growth, conveniently overlooking the promiscuous relations that the regime has with the foreign investors, bankers, and trading companies.

Labor policy: 'reforms' that benefit the bosses

What is important in analyzing a political leader is not where he comes from, but where he is going, not his past cohort, but his present and future reference groups. Political observers have been wrong in their analysis of Da Silva because they focus on his distant past, his former trade union comrades, not his present allies among neoliberal bankers, businessmen, and imperialist regimes. When Da Silva proposed a social pact between labor, business, and the government purportedly to work for the betterment of the country, he set up a Social Economic Development Council to formulate policy recommendations. The composition and agenda of the Council revealed Da Silva's pro-business, anti-working-class bias. Of the 82 members of the Council, 41 are businessmen (including Gustavo Marin, head of the US financial giant Citygroup) and 13 are trade unionists, a better than three to one ratio in favor of the bosses. The purpose is to discuss tax 'reform,' reduce business taxes and social security, and cut payments to workers, pensioners, and other beneficiaries. When Da Silva was confronted with the preponderance of the business elite, he roundly defended his pro-business bias, embellishing his choices with an apolitical, meritocratic varnish and accusing his critics of nepotism. 'This Council,' Da Silva argued, 'is not a friends' club. I am not interested in knowing the party affiliation [sic] of the members of the Council or for whom they voted. What interests us is the competence, the capacity, their talent and knowledge to think for their country' (*Tiempos del Mundo*, Dominican Republic, February 20, 2003, p. 7). Lula here conveniently forgot that his businessmen's disinterested talent for thinking for the country has resulted in the greatest social inequalities in the world. Da Silva deliberately overlooked the class interests of the business elite precisely because they are his strategic allies in his pursuit of orthodox neoliberal policies. The rightwing alliances of Da Silva have already enmeshed his regime in a major scandal. In late February it was proved that rightwing Senator Antonio Carlos de Magallaes from Bahia had tapped the telephone of over 200 congressmen, senators, and other prominent political figures. The senator supported Lula during the presidential campaign and was seen as a strategic ally in providing support for his neoliberal legislative agenda including labor 'reform.' When numerous deputies demanded congressional hearings, President Da Silva and his inner nucleus of advisors ordered PT congresspeople to vote against the congressional investigation, badly tainting the image of the 'honest and open people's president.'

Da Silva's labor reform strategy is directed toward weakening the trade unions, undermining constitutional guarantees of labor rights, and lowering labor costs to increase profits for employers under the guise of making exporters more competitive. His legislation proposes to eliminate payments by private sector capitalists to the trade union funds and abrogating obligatory payments of union dues. His second piece of legislation proposes to allow capitalists to secure labor contracts that override legally established workers' benefits (*Financial Times*, November 26, 2002). The former metal worker bashes his workmates and repays the CUT for its electoral support by topping his legislative agenda with the principal demands of the industrialists' association.

Lula's mechanism is to co-opt the bureaucratic bosses of the CUT by offering them positions and stipends as advisors to his regime. CUT president Joao Felicio, one of the co-opted bureaucrats stated 'We [sic] have a certain sympathy for the reforms, but they have to be negotiated and imposed gradually.' The trade union national secretary of the PT, Hergurberto Guiba Navarro, bluntly stated the purpose of labor reform: 'We are going to undertake a grand reform and many unions will disappear' (*Financial Times*, November 26, 2003, p. 8).

Given Lula's hard push of ultra-neoliberal orthodoxy and the cooptation of the CUT leaders, it is not surprising that working-class opposition comes from the public employees union, dissident CUT unions, and, to a lesser extent, the rightwing trade union confederation Forza Sindical (FS). In March, the metal workers affiliated with FS went on strike over declining real wages. FS is making a show of fighting to reduce the working week from 44 to 40 hours, to increase severance pay and extend unemployment benefits (to increase coverage from five to twelve months), and for legal recognition of workers' representation on the shop floor. The regime is adamantly opposed to all FS demands, claiming they are inflationary and threatening repressive measures against what they dub political demands, an old ploy employed by all previous rightwing regimes, prior to bringing down police batons on striking workers' heads.

Human rights

The level of human rights violations is directly related to the economic strategy adopted by governments. Throughout Latin America regimes intent on dismantling social welfare legislation, lowering living standards, and especially promoting foreign investment and primary commodity exports for highly concentrated elites have a notoriously bad record on human rights. This goes for military or civilian electoral regimes intent on sacrificing labor to provide incentives to foreign investors and local financiers. The regime's record on human rights is a prime illustration of this hypothesis.

The most rigorous and systematic collection of data on human rights violations in the Brazilian countryside is conducted annually by the Pastoral Land Commission (*Comissao Pastoral da Terra*—CPT). Our discussion of the violations of human rights during the Da Silva regime will be based in large part on CPT's findings for 2003. We will then analyze and discuss human rights in relationship to the economic strategy and its implications for understanding the politics of the regime.

There are several measures to evaluate the regime's human rights record during their first year in office. These include (1) the assassination of activists; (2) the jailing of peasant leaders and social activists; (3) the activities of the paramilitary groups; (4) the impunity of the military; (5) equal protection before the law; (6) recognition of the legitimacy of the land reform movement; (7) the end of forcible eviction of landless squatters on uncultivated land; and (8) realization of extensive land reform.

Paramilitaries at the hire and service of the landlords have expanded throughout the rural regions of Brazil during the first year of the regime. The paramilitary forces operate with impunity, their presence has been televised and their interviews have been broadcast on the national media. In Parana, Para, Bahia, and throughout the northeast, north-central, and even southeast of Brazil the paras frequently operate in association or complicity with the military police and with the tolerance of the judiciary. These private security forces have assassinated the great majority of peasant leaders under Lula's 'hands-off' policy.

The CPT's national campaign to outlaw these armed militias has elicited broad support from Brazilian and international human rights groups. This has had virtually no impact on the Da Silva regime, which argues that under the separation of powers this is a 'judicial affair' to be handled by the 'states.' Lula's policy has led to the proliferation of new paramilitary groups and death squads, including the First Rural Command *(Primeiro Comando Rural)* in Parana, which have targeted over 14,000 land settlement families for eviction.

In September 2003, 150 military police surrounded the MST headquarters in São Paulo and prepared for an armed assault on the pretext of looking for social activists accused of property violations (land occupations). Only the massive intervention of human rights groups, the Catholic bishops, and trade unions prevented a potentially bloody assault. The Da Silva regime eventually relented to avoid further tarnishing its international image with a massacre in Brazil's largest city. No inquiry was launched, nor were any officials reprimanded—and of course there were no 'criminals' in the headquarters. The media effect, however, was to criminalize the social movements in general and the MST in particular.

Da Silva, who actively sought and received the wholehearted support of the MST and social movements during the election campaign, has washed his hands of responsibility for the growing judicial persecution, the arbitrary arrests, and intervention of military police. Claiming 'the separation of powers' between the executive, legislative, and judiciary, he has refused to use the authority and influence of his office to call off the forces of repression or uphold constitutional guarantees against arbitrary arrests and extrajudicial executions by paramilitary groups linked to big landlords. The reason for Lula's unwillingness to act is found in his deep commitment to the promotion of the agro-export model, to sustaining a 'favorable climate' for foreign investors and his perception that any interventions against big capital and their judicial, police, and paramilitary allies would send the 'wrong signals' to the 'market.'

It ill behoves a president who has supped in the houses of landless peasants to claim 'neutrality' in this fundamental struggle for social justice and human rights. Da Silva's policies however are not 'neutral,' whatever his claims and recitation of textbook clichés about the division of powers. In effect, his policies have given license to the most retrograde forces among the Brazilian elite to *roll back* the gains achieved over the past two decades by encouraging the eviction of land reform beneficiaries and land squatters, and to fomenting the lawless behavior of landlords and the corrupt judiciary which acts at their behest.

At the end of her three-week visit in September and October of 2003, the UN envoy, Asma Jahangar, investigating summary executions by Brazilian police, noted that 'Brazil is a democracy. But what I see here is a wretched, sad situation where there is no justice' (BBC News, October 10, 2003). Two of the witnesses who gave testimony to the UN envoy on the operation of death squads in rural and urban areas were murdered shortly thereafter as if to confirm the pathetic state of human rights in Brazil. The UN representative noted that the problem is not merely a few local vigilantes but an institutional problem that permeates the Brazilian state. The UN investigator gathered detailed and extensive reports from human rights groups that linked the death squads with police officers and vigilantes. As Jahagar noted, 'The police cannot fight crime by committing crimes.' Da Silva has paid lip service to the problem, but has failed to undertake any serious attempt at instituting reforms of the police, judiciary, or other institutions concerned with law enforcement. In effect, the 'law' of immunity flourishes under Da Silva as it has under previous military and civilian regimes.

Brazil has the most extreme inequalities in land ownership in the world. Less than 1 per cent of the landlords own 50 per cent of the land, while 25 million rural families are landless. The question of agrarian reform was the central demand of the most impoverished rural classes in Brazilian society, a demand backed by over two-thirds of the Brazilian public. Da Silva, during the election campaign, promised a 'profound, integral agrarian reform within the law.'

From January 1 to November 30, 2003, the CPT counted 71 assassinations of rural workers, an increase of 77.5 per cent over the previous year (43 assassinations in 2002) and the highest since 1990. In addition, there was a 76.3 per cent increase in attempted assassination (67) over 2002 (38). Serious injuries doubled in 2003 from 25 in 2002 to 50. The number of political prisoners increased from 229 to 265 in 2003.

Under the Da Silva regime there was a 227 per cent increase in families ordered off the land by judicial order, while the number of landless squatters forcibly removed rose by 87.8 per cent over the previous year. The year 2003 established a record in judicial expulsions: 30,852 families were served with 138 judicial orders by military police to abandon the land they were seeking to cultivate, the highest number in nearly 20 years of record-taking by the CPT. In 2002, 63 judicial orders were issued evicting 9,243 families. A conservative estimate of four persons per family would mean that over 120,000 persons were turned off the land and onto the highways. In addition, there was a sharp increase in families forcibly evicted by gunmen and local landlords without judicial authority. In 2003, 2,346 families were expelled compared to 1,249 in 2002, an increase of 87.8 per cent.

The Da Silva regime hypocritically claimed that the federal executive power could not intervene, because these were matters for the judiciary, because there was a separation of powers, that these crimes against rural workers were outside of federal jurisdiction, and so on. In fact, as president he is charged with upholding the constitution; he has the constitutional authority to support land occupations of uncultivated land, as stated in the Brazilian constitution. Da Silva's tolerance, if not overt complicity, in the violent repression of rural workers strongly suggests that in the increasingly polarized countryside he has taken the side of the big landlords.

There are several factors explaining the Da Silva regime's benign neglect of human rights. First and foremost since the big landholders involved in the agro-export sectors are strategic actors in the regime's policy of surplus generation to meet debt obligations, Da Silva is very reticent about becoming involved in conflict involving any sector of the big landlords that might 'unsettle' the big agro-exporters. Hence Lula's resorting to the subterfuge of 'limited jurisdiction' and the 'separation of powers.'

Da Silva's claim to 'limited powers,' however, does not apply when it comes to positive action on behalf of the landlords, as is seen in his privatization-by-decree of state banks (bypassing Congress) and his more than two dozen trips abroad promoting business for agro-exporters. The sharp increase in human rights violations under the Da Silva regime can also be accounted for by the increase in the number of land occupations by the rural landless movements. Under the mistaken assumption that Da Silva was a 'friend of the movements' and that he would fulfill his promises to carry out a comprehensive land reform, the MST and other rural movements increased their activities, believing that they were helping Lula to realize his promise. Faced with increased activity from the landless movements, and encouraged by Da Silva's promotion of agro-business and his unwillingness to enforce the agrarian clauses of the Constitution, the big landholders called in their private armies (dubbed 'private security forces'), their corrupt allies in the judiciary, and local and state police to evict thousands of families forcibly. These violent activities were frequently preceded or followed by selective assassinations of activists.

In line with Da Silva's general policy of neglect vis-à-vis human rights violations, landlords were emboldened to increase the use of slave and child labor, both of which are rife in the countryside, notwithstanding the government's support of UN (ILO) prohibitions and warnings in regard to these forms of abusive labor practices. There was a sharp increase in the number of cases of exploitation of forced labor and the number of slave workers. In 2002 there were 147 cases, compared to 223 in 2003, an increase of 51.7 per cent. The number of slave workers in 2002 was 5,559 compared to 7,560 in 2003, a 35 per cent increase. The state of Para had the highest number of denunciations of the use of slave labor (169) involving 4,464 workers. Only 40 per cent of the slave workers (1,765) were liberated.

Similar data have been reported on the growth of child labor in part because of the lax enforcement of the Da Silva regime, its low priority to social issues in relation to foreign investors, the sharp increase in unemployment, and the abandonment of tens of thousands of landless workers encamped in plastic tents.

Da Silva's reticence to use his federal powers to enforce human rights is not a problem of law or constitutional constraints; it is a question of politics. To pass his orthodox economic policies and his regressive social policies, Da Silva has formed strategic alliances with rightwing parties and leaders. These politicians have long-standing ties to big landlords and the corrupt judicial officials who commit human right violations. Da Silva is thus indirectly allied with the most retrograde sectors of the rural elites, the major socioeconomic bases of his political allies in the Congress and Senate. This alliance with the traditional right was evident in the congressional vote on the constitutional amendment to slash the pensions of public employees drastically. In 2004, Da Silva's alliance with the right became evident at the cabinet level with the inclusion of the conservative Brazilian Democratic Movement in the cabinet.

Even the liberal press ridiculed the extreme measures which the Da Silva regime went to meet the harsh demands of the IMF. On Christmas Day 2003, *Folha de São Paulo* published an article pointing out that the government withheld spending over a quarter of the budget allocated for the Fund to Combat and Eradicate Poverty in order to achieve the budget surplus agreed to with the IMF. Of the $1.7 billion allocated for poverty reduction, $430 million was subtracted to exceed the budget surplus destined for foreign and domestic creditors. With $430 million, the regime could have easily settled 100,000 landless rural workers, provided food for ten million hungry children, or halved the 21 per cent unemployment rate of greater São Paulo. Few if any of the recent and plentiful crop of orthodox rightwing, neoliberal regimes have manipulated budgets to such an extent to 'over-fulfill' IMF goals. The denial of available allocated public funds to millions of hungry, unemployed, and landless Brazilians is probably among the worst human rights offenses of the regime, a gratuitous insult to the poor through excessive servility to the IMF.

Zero hunger, zero accomplishments

With his usual theatrical demagogy, Da Silva proclaimed early in his presidency that all Brazilians would have three meals a day by the end of his term. He then journeyed to his home town and announced a campaign of 'zero hunger,' a program of providing a food basket to each and every family suffering from hunger. The program was a total failure, from every angle. To start with, the initial program was cut by $10 million to adjust the budget to pay the affluent creditors. Secondly, the allocations of food reached just a tiny fraction of the hungry. Thirdly, the increase in unemployment and absence of agrarian reform increased hunger beyond the meager handouts of the top-heavy and underfunded and rather inefficient bureaucracy running the program. Even in the 'showcase' town where Da Silva was born the program failed. At the end of December 2003,

Bishop Irineu Roque Scherer, whose jurisdiction in Pernambuco State includes Caetes, Da Silva's home village, declared, 'The people believed, despite the drought, that with the election of a countryman as president they would secure water, but nothing has changed up to now.' Caetes is administered by Da Silva's party, but thanks to the local and federal governments' inaction and neglect the small farmers lost 90 per cent of their crops of corn and beans. The bishop pointed out that Da Silva 'has a pretty speech that charms and convinces local people but the PT doesn't follow through. Consequently the government promises, but nothing happens' (*Jornal de Estado de São Paulo*, December 31, 2003).

Cardinal Paulo Evaristo Arns, a long-time personal friend of Da Silva, criticized the regime for its 'indifference' and hoped that in 2004 Da Silva would be more 'realistic' and not simply be a man of all talk. He went on to state that 'perhaps the [economic] situation for many people has gotten worse, given the promises he made' (ibid.). Even minimum investments in excavating wells for irrigation were not undertaken in the northeast according to Bishop Irineu. The president of the National Conference of Bishops, Geraldo Majella Agnelo, pointed out that food baskets were not the answer to poverty, 'Agrarian reform is the most important reform which the government can carry out because it goes to root of social injustice.' According to the bishops, Da Silva's 'zero hunger' has not gone beyond emergency, piecemeal activities and has failed programmatically. Even in his home town, the food baskets have reached less than a quarter of the 2,000 needy families living on R55.8 per capita a month (less than $20).

Even if Lula had carried out his food handout program to the 40 million hungry Brazilians it would have amounted to $10 a year, 85 cents a month or 2.5 cents a day, per person, enough for a banana for each family of five. Lula, the self-styled 'workers' president,' had the self-proclaimed 'courage' to raise the budget allocation to meet debt obligations from $17 billion to $19.4 billion, an increase of 14 per cent. The 'people's president's $2.4 billion addition to debt payments was a direct transfer from the social budget. His reputation as a 'Robin Hood for the rich' is well known in international financial publications. On the last day of 2003, a journalist from the *Financial Times* (December 30, 2003, p. 12) graphically described the impact of Da Silva's economic policies:

A year ago Joao Baptista Andrade took his son on a 16-hour bus ride to Brasilia to celebrate the inauguration of Luiz Inacio Lula da Silva as Brazil's first working-class president.

These days Mr. Andrade spends much of his time in a long unemployment line in downtown São Paulo. 'Of course I'm disappointed. Lula promised us 10m jobs in four years and today there are fewer than a year ago,' he says.

Across town at Daslu, an elite fashion store where the rich drink tea while trying on designer shoes at $1,500 a pair, business is flourishing. A customer exits with half a dozen servants carrying her holiday purchases.

'Lula seems to have come to his senses, I thought I was going to have to move to Miami,' she says before disappearing into her air-conditioned limousine.

Lula's 'Robin Hood for the Rich' policies led to astronomical 130 per cent returns for speculators and investors. A leading business magazine declared Lula's Minister of Finance Palocci 'Man of the Year.'

A detailed study of Lula's 'zero hunger' program published in *Outro Brasil* in November 2003 reveals the weaknesses and ultimate inadequacies of the program. The researchers divide the program into two: emergency aid food baskets for the indigent (receiving less than half the minimum wage) and structural changes. According to the study, only 6 per cent of the budget was spent in the first quarter of the year. The projected population to be targeted if funds were available was limited to less than 10 per cent of the 'hungry' for brief intervals of time. The emergency food program was neither comprehensive, sustainable, nor adequately supervised and coordinated by the competing bureaucratic agencies. None of the structural changes related to hunger—employment, income, agrarian reform, irrigation, and so on—was pursued.

The distribution of 'hunger funding' is very similar to the 'food baskets' which the traditional party bosses hand out to the poor in sustaining their electoral machines, and neutralizing opposition to the unjust land tenure system. The problem of 'food for the hungry' is ultimately related to the 'free market' macroeconomic policies, which result in subsidized food imports destroying local food producers and increasing rural poverty far beyond the paltry handouts of the Da Silva regime.

Taliban neoliberalism and the rights of Indians

At the end of December 2003 several thousand Indians invaded a number of farms close to Brazil's border with Paraguay to reclaim land stolen from their families. As the Guarani and Karowa peoples occupied two large plantations in the state of Mato Grosso do Sul and camped outside several others, the landlords fled. All the ranchers in the region illegally occupy Indian lands, setting up large-scale cattle ranches and forcing the Indians into destitution and many to suicide. While these conflicts have been simmering for some time, the Indians like all the popular sectors had received assurances from Lula that their claims and rights would be respected. After waiting twelve months and experiencing no substantial results from Brazil's FUNAI (National Indian Foundation), the Indians resorted to direct action. FUNAI officials in turn claim that the regime has cut funding, personnel and lowered regulations, thus undermining their capacity to act on behalf of the Indians.

Throughout the Amazon, Indian communities are bitterly criticizing the regime for not acting to control the growing invasion of lumber barons, plantation owners, gold miners, and landless settlers into their lands. In the state of Rondonia, the governor and the local judiciary work hand in glove with criminal gangs exploiting diamonds in the territory of the Cinta Larga Indian community. The killing and infection of the Indian population continue unabated. Over the past 30 years the Indian population has been reduced by a third. In the recent past, 2.7 million hectares of Indian territory of the Cinta Larga has been illegally exploited and the land degraded by lumber companies exporting logs to the US and EU.

The Da Silva regime has not taken any measures to enforce Indian rights. On the contrary, its emphasis on exports and hard currency has encouraged more exploitation and greater incursions. The budget cuts have, according to FUNAI officials, undermined any effort to defend Indian rights. Lula's new political allies include the rightwing governors and politicians promoting and defending the predators of the Amazon and the Indian territories, and that is probably the main reason the regime refuses to act. Lula prizes the rightwing votes in Congress in favor of his IMF agenda over the rights of the Amazonian Indians. Indian resisters are arrested, prosecuted, and sentenced to jail; the crimes against Indians are not investigated. Gunmen for the timber and plantation intruders murdered 23 Indians in Brazil in 2003, including five in the south. In the meantime, hundreds of Indians have been physically assaulted or threatened, including two school buses transporting Indian school children. Impunity and unsolved crimes are the norm under Da Silva's administration as it was under his predecessors who permitted big capital and their political backers at the local and national level to devastate the Amazon and continue their historic policy of genocide against the Indians.

Social movements and trade unions

The CUT, the left labor confederation with close ties to the PT, particularly Lula, had numerous leaders elected to Congress and some are ministers in the regime. So far few, if any, have voiced criticism of Da Silva's right turn. The CUT itself, though claiming 15 million affiliates, is largely bureaucratized, with a large staff and dependent on state funding. CUT's power of convocation is very limited; no more than a few thousand turn out for major protests. From the beginning of the Da Silva regime, the CUT leadership has adopted a double discourse. Shortly after Da Silva's election, the CUT was invited to discuss the new regime's 'social pact' to reduce pensions, postpone wage and minimum wage increases, and weaken the financial basis of trade union funding. The CUT leadership declared its independence from the government but agreed to continue to participate in the Social and Economic Council even though businesspeople and bankers outnumbered the trade unionists by more than three to one. Subsequently, the CUT continued to criticize the harsh neoliberal budget cuts and reactionary reallocation of funds favoring local and foreign bondholders, while continuing to support the regime. Worse, on the purported rightwing Social Pact the CUT's main difference with the neoliberal economic team was the manner of its implementation—advising the neoliberals to 'gradually implement' the anti-working-class measures, rather than imposing the whole packet of harsh measures immediately. The CUT's servility to the regime is a continuation of the negotiating posture it has adopted with previous neoliberal regimes, in part because of its dependence on government subsidies. In addition, there are strong structural ties to the PT via the ex-CUT officials serving in the regime and the promise of a future position in the government or inclusion on the list of deputies for the next congressional elections. Finally, there is the bureaucratization of the CUT; its leaders and staff have been running the unions in vertical fashion for over a decade, marginalizing militants, and are totally incapable of organizing the vast army of unemployed and underemployed. The results were evident in any major protest demonstration regarding ALCA, IMF, or the rash of privatizations under Cardoso. The CUT leadership, having demobilized its membership for over a decade, was not able to put more than a few thousand in the street, and most of those CUT members present were largely mobilized by militants from PSTU, PC do B and the left wing of the CUT. Leaders of the MST have told me that the progressive sectors of the Catholic Church can mobilize more people than the official leaders of the CUT. What confuses outside observers of the CUT is the fact that its leaders show up to make speeches or sign declarations in favor of radical demands, giving the impression that it is still a radical mass trade union. Despite the harsh anti-labor legislation envisioned by the Da Silva regime, there are few signs of active opposition from the official leaders, though by late December 2003 many class-conscious trade unionists were shocked and angry at what they perceived to be Da Silva's pro-business partisanship. Some investment advisors give Lula another year before major conflicts break out to challenge his neoliberal agenda, urging Palocci and the rest of Da Silva's team to take the fast track and ram the 'bitter medicine' through Congress before the poor, the landless, and the trade unions overcome their illusions about the 'people's president.'

The left intellectuals

The intellectuals backing the regime can be divided into orthodox Lulistas and the neo-Lulistas, attracted by the neoliberal policies and the chance to secure advisory positions in the regime or state 'contracts.' The main role of the orthodox Lulistas, at least in the first six months of his regime, was to polish Da Silva's image as the 'people's president,' apologize for his regime's reactionary pro-imperialist policies by citing the 'difficult and complex world,' 'the impossibility of breaking with the IMF now' and elaborating a new 'pragmatic' approach, which seeks to balance Lula's rightwing economic policymakers with the so-called 'leftists' who operate in the interstices of tight budgetary and ideological restraints of the dominant rightwing group. Those intellectuals who sang to the tune of the FSM's 'another world is possible' now add a new refrain, 'not, now, another day it's possible.' The new pragmatists also serve as the ideological hatchet men who disparage and dismiss leftist critics of Lula's rightwing policies.

The neo-Lulistas are not as harshly critical of the leftist critics, as they do not feel any obligation to cover their tracks to the right. By beliefs and practices, they position themselves as 'technocrats' and 'progressive' neoliberals who are interested in a 'heterodox' free market model that combines competitive markets and social spending, though they spend most of their efforts on the former and usually point to the future regarding any obligations to what is dubbed the 'social debt.'

The left intellectuals are spread across the political spectrum. Some remain part of the left of the PT, others are outside the party as well as the regime. A group of critical intellectuals have joined the expelled former deputies of the PT to build a new political and social movement to contest the savage cuts adopted by Lula. Their party, 'Socialism and Democracy,' proposes to combine support for popular and electoral struggles. A great number of intellectuals who hoped that they could influence the regime via the progressive ministers in the government or via outside pressure have become pessimistic and demoralized.

By the end of year 1 an increasing number of intellectuals began to think in terms of regime change. But as yet their political future is undetermined. Some are drawn to the new party, others to the Marxist PSTU, but a few have taken an independent critical course.

The MST

The Rural Landless Workers Movement faces a profound dilemma: after years of building a successful mass independent socio-political movement that settled over 350,000 landless families on unproductive land via direct action (land occupations), it temporarily substituted electoral work for Lula in the hope of positive agrarian reform legislation. They have been sorely disappointed. The past success of the MST was based on its capacity to prioritize independent mass action, even as it supported some progressive electoral candidates of the PT. Having relied on Lula's election as the fulcrum for a comprehensive agrarian reform, they are now faced with a regime that has repudiated every one of their supposedly 'shared reforms.'

For several years prior to the presidential elections, there were open debates and discussions in the MST regarding the movement's political future. Some argued that the PT was turning into a conservative or social democratic electoral party and that many of their state and locally elected leaders were hostile to agrarian reform and, in some cases, actually repressed land takeovers. They concluded that the MST should form its own party with other social movements and leftist groups. A second group conceded that the PT was becoming more conservative and they also repudiated the rightwing PT governors and mayors, but they argued that the MST should run its own candidates in the PT, or at least work more actively within it to influence it in a more progressive direction. The third force and the most influential, at least among the national leadership, tried to bridge the differences. They agreed to work outside the PT to try to coalesce with the progressive church, human rights groups, and left intellectuals to elaborate an alternative program and organization. Thus was born the Consulta Popular (CP), which began with a great fanfare and then declined in part because it was combined with the old tactic of influencing the PT from within. In effect, the CP was neither a new movement nor a new electoral party. It was squeezed between direct action and electoral politics and was not able to attract any sizeable trade union or urban support.

Lula's electoral campaign of 2002 demanded and secured from the MST an unprecedented concession: the stoppage of all mass direct action—no land occupations—arguing that this would 'play into the hands of the right,' 'scare' middle-class voters, and cost Da Silva the elections. For the first time MST fell into the trap. They called a halt to mass action and joined the electoral campaign despite Lula's reactionary alliances and the clear hegemony exercised by pro-imperialist interests. The MST substituted vague 'populist' statements for class analysis. After all, tens of millions of poor would vote for Lula and their expectations for a rupture with neoliberalism would force him to respond positively.

Predictably, on taking office, Lula ignored the 'popular expectations,' or better still asked 'forgiveness' for ramming the neoliberal pole up the populace's backside. Unfortunately, most of the MST leaders continued to hold out hope that Lula and the impotent Minister of Agrarian Reform, Miguel Rassetto, and other left functionaries in the same ministry would make a 'left turn.' Rossetto, a member of the left social democracy tendency of the PT, argued he would do everything to comply with the agrarian reform promises within the extremely limited budget constraints imposed by his government—a clever piece of demagoguery.

Meanwhile, tensions were mounting within the MST as rank-and-file activists and over 200,000 land squatters who were camped out under plastic tents, suffering heat, cold, food shortages, and mosquitoes were becoming restless. An increasing number of land takeovers began to take place. A movement like the MST must act or disintegrate. No positive measures were forthcoming from the regime. Agrarian reform was relegated to the back burner, along with zero hunger and other of Da Silva's electoral promises. The argument of some MST leaders to work from within was wearing thin. Some national and regional leaders publicly expressed their discontent with the government's unresponsiveness (*Folha de São Paulo*, February 9, 2003). The government appointed several progressives sympathetic to the MST and other groups to the Agrarian Reform Institute (INCRA) but with few resources. A few later resigned or were fired. More important, Lula has taken an extremely rigid and hostile position on the traditional land occupation tactics of the MST, a promise to apply the full force of the law to repress the movement. He argues that any agrarian reform measures will have to be part of a regime-sponsored program, which the post-election budget promises to be totally insignificant.

Many of the regional and local MST leaders and activists recognize that the landless rural workers have no future with Lula's regime, that the movement will have to part company and return to the tried-and-tested method of mass direct action.

A threat to the left

The Da Silva regime represents two dangers. In the first instance, the regime represents a threat to the living standards, working conditions, and social life of the vast majority of small farmers, landless rural workers, wage and salaried workers, and pensioners of Brazil. The threat is all the more acute because it comes from political parties or a coalition of parties and social organizations that were the prime defenders of the working and peasant classes, who have now joined their enemies and thus leave the masses temporarily defenseless. In addition to the physical pain and social suffering which the Da Silva regime is bringing about, the right turn will cause immense social psychological damage, provoking mass disillusionment not only with the PT regime and its public faces, but also mass disenchantment with the whole spectrum of parties, trade unions, and social movements that promoted Lula as the 'people's president.' Equally important, the PT ideologues, like Frei Betto, who have justified Lula's politics as 'realistic' and 'pragmatic,' have made plausible, especially to ill-informed leftist intellectuals, the idea that there really is no alternative to adapting to the neoliberal policies. By assimilating Lula's rightwing policies to a general leftist label, the Lulista ideologues threaten to redefine the left along the neoliberal politics of the Spanish Socialist and British 'New Labour' parties, in effect, emptying Brazilian leftist politics of its essential welfare and socialist content.

Secondly, the international left, which has joined the Lula chorus, is leading the popular movement toward a massive political debacle. The ill-informed effusive praise of Lula's electoral victory as the greatest revolutionary change since the Cuban revolution, Salvador Allende's election or the Sandinista revolution, prepared the ground for a popular disillusionment as the reactionary policies begin to penetrate popular consciousness.

Two outcomes are likely. On the one hand, a part of the Latin American left will take Lula's rightwing path as a model and abandon historical anti-imperialist and redistributive popular demands, citing the 'constraints' facing Lula and other such rationalizations. This course has been adopted by the Broad Front (*Frente Amplio*) in Uruguay. On the other hand, other leftwing movements will rethink the entire electoral strategy, particularly the relation between party and movement. From a practical, historical perspective, it is clear that the divorce of the PT from the mass movement and mass struggle early on laid the groundwork for its class collaboration practices and eventually pro-imperialist regime policies.

The dynamics of class struggle and the emergence of direct action mass movements like the MST were instrumental in creating a challenge to the neoliberal orthodoxy, particularly in the context of failed neoliberal states. Economic stagnation, deepening inequalities, ballooning external debt—together with a leftist critique—created the basis for the decline of the traditional neoliberal right, but not the sufficient conditions for the rise of radical or even reformist alternatives. Instead, the political conditions of a new virulent orthodox neoliberalism has emerged based originally on the working class, the middle class, the landless workers, led by plebeian ex-leftists, but now allied with and subordinated to the agro-mineral elites and finance capital.

The PT's radical rupture with its leftist past did not result in losing popular support in the short term because of the plebeian nature of the leaders, the manipulation of popular imagery, and the hierarchical, personalistic, and authoritarian nature of the party leadership. The popular origins of leaders neutralized internal opposition and enforced conformity with the rightwing course during the first year. After all, who was willing to stand up to the 'people's president' when Lula embraced George W. Bush, the eminent warmonger of our epoch, and called him an 'ally of Brazil?' Who stood up among the pragmatic ideologues of the 'people's movement'?

Lula has a clear, coherent, neoliberal strategy based on an alliance with the IMF, Washington, overseas investors, and creditors. He and his advisors have put in place an effective strategy to limit internal party opposition, using the carrot (of offering ministries and secretariats) and the stick (threats of censure and expulsion to persistent critics). Through state patronage and party discipline, he has converted PT mayors and members of Congress into transmission belts for his harsh austerity programs. There are exceptions, of course; a handful of PT elected officials who still uphold the traditional social democratic, reformist program, but they have been marginalized, abandoned in large part by their former comrades with a voracious appetite for the spoils of office and small fiefdoms of state power. The regime has the power and will to impose the harsh neoliberal policies on the country on the lower classes and enforce compliance within the party.

The overwhelming control of the PT leadership was manifest in the first meeting of the National Directorate after Lula's election on March 16, 2003. Two proposals were put up for approval. The neoliberal resolution supporting the rightwing political economic course of the regime received 70 per cent of the votes (54 votes), the left dissident proposals received 28 per cent (21 votes), and there were two abstentions. The resolution explicitly established in doctrinaire fashion the arguments and logic justifying the policies of the regime, establishing the theoretical and practical reasons for the adoption of the neoliberal strategy (monetarism, adjustments, etc.). The resolution affirmed that the pro-business policies and support for the IMF were not *tactical* but principled positions. The meeting also reflected the consolidation of control of the party apparatus and the almost total marginalization of the left tendencies. The resolution, the meeting, and the vote left little doubt that there was absolutely no hope of reforming the party from within, or pressuring the leadership to make a 'left turn.' Staying in the PT means supporting the party of the IMF, George W. Bush, ALCA, the enemies of Venezuela's President Chavez and joining border patrols with Colombia's paramilitary president Uribe—an indefensible position, at least from a popular leftist perspective.

Da Silva's opposition in the PT, in contrast, is ideologically disoriented and strategically and tactically impotent. Unwilling to embrace Lula's radical 'redefinition' of the 'reformist' program (from social welfare to orthodox neoliberalism), they search for a new strategy and program. Outside of the PT, some of the movements have narrowed their horizons, setting aside their opposition to Lula's general embrace of the pro-imperialist agenda in favor of seeking 'sectoral reforms'; agrarian reform, urban programs for the *favelados*, etc. Even in these 'sectoral strategies,' the opposition has scaled back their demands in their effort to 'realistically' adapt to Lula's budget cuts and compliance with foreign creditors.

In the first year the opposition left PT and the social movements, having put all their efforts into supporting Lula, continued the hopeless task of working within the elite, hierarchical party apparatus. They had no influence in changing the course of the regime.

What about an 'outsider strategy,' the approach of those who have decided to oppose the regime from the outside? These include the new Socialism and Democracy Party, the PTSU, and others. Strategically, they should be in a powerful position. Lula's regime and its neoliberal policies will lead to a more profound social, financial, and economic crisis than that which affected the Cardoso regime. Budget cuts and the payment of debt undermine productive investments, weaken the domestic market, increase future debt obligations, and lead to stagnation.

The decline in pension payments, the real reduction in the minimum wage, and the deterioration of social services together have lowered living standards and income by 15 per cent. Payments to wealthy bondholders, subsidies to big agro-exporters, and inflation widened inequalities. The extreme right turn of the regime, the precipitous decline in living standards, and the deepening recession will eventually lead to a decline of Lula's initial high popular ratings.

The PT's expulsion of three dissident Congress members and one senator led them to create a new political party. The four have substantial national and regional backing. Early in 2004 they convoked a national meeting to form a new party. More seriously, there are differences between the new party and the PSTU over the nature of the new party and it appears that they will not be able to combine forces.

The question of the effectiveness of the leftwing political opposition to the Da Silva regime is crucial. The small but disciplined United Socialist Workers Party (PSTU) has been gaining influence among trade union militants in the CUT and currently influences about 10 per cent of the Confederation. The PSTU and the Socialism and Democracy Party have potential for growth, but can become a formidable opposition only if they find allies with more significant social movements, church dissidents, and trade union forces. One such configuration could find leftwing MST leaders, a sector of the CUT, progressive Catholic clergy, and left parties coming together to form an alternative opposition coalition or political party, one that focuses on mass direct action over and against electoral politics. This possible formation has tremendous potential in taking up the banners of anti-ALCA, debt repudiation, internal market development, agrarian reform, and renationalization of strategic industries and banks. Millions of Brazilians are on record in opposition to every one of Lula's policies. The anti-ALCA referendum was supported by ten million voters; of the 52 million who voted for Lula, the overwhelming majority voted for a political-economic rupture with the past neoliberal policies, not a continuation and deepening of the same.

Despite the favorable strategic objective and even subjective conditions for the re-emergence of a new leftwing formation, there are several severe limitations. First is the absence of a mass social movement with a national presence that is capable of serving as a pole for regrouping. The new mass political party has to be created in the social struggle. First, it will be led by social and political fragments of the exploited classes. Secondly, the new political formation will have to engage in a harsh ideological struggle to unmask the 'people's president' and expose the profoundly reactionary, continuist nature of his regime. This will take time and effort because the defenders of the regime range from the majority of the mass media to the polemical ideological apologetics by ex-leftists associated with the da Silva regime. Thirdly, the new political formation will have to achieve a high degree of principled political behavior, avoiding alliances with rightwing critics, though there is plenty of room for possible tactical alliances with the moderate trade union, *Forza Sindical*, on issues of wage, salaried, and labor legislation. Fourthly, the political formation must develop theoretical and programmatic clarity, regarding the nature of the neoliberal crisis, the new militarist colonial imperialism of the US, and the major contradictions undermining the viability of Lula's economic model.

Finally, the new political formation must organize and organize and organize. More than 90 million Brazilians live in poverty, most of whom are not organized and will be further impoverished by Lula's policies, the so-called zero poverty program notwithstanding. There are 25 million landless Brazilians in the countryside, 95 per cent of whom will not benefit from any land reform, but will be further marginalized by Lula's promotion of agro-export strategies. There are

40 million under- and unemployed who have no future employment prospects, given Lula's budget cuts and high interest rates. Hundreds of thousands of small and medium-sized enterprises (and not a few large national firms) face bankruptcy from the high cost of credit and the free trade policies (ALCA) promoted by Lula.

The political opposition has a formidable challenge in organizing the unorganized, otherwise there will be spontaneous protests which will be harshly repressed as Lula has promised the international investor class. They will have to face mass disenchantment that could be attracted to the rightwing patronage parties who support Lula today, but who will abandon a sinking ship, as they have always done in the past.

Finally, the new political formation, while appealing to the discontented voters abandoning Lula, must make a thorough and complete break with the PT, a party which, like many others in Europe and Latin America, began on the left and has become the New Right. There is no inevitable outcome to the Brazilian experience. Objective conditions are favorable, subjective opportunities are emerging, but the question of political leadership is still open.

Conclusion: perspectives for 2004 and beyond

Regime economists and the international financial institutions (IMF, WB, and so on) are predicting that Brazil will grow by 3.5 per cent in 2004 based on a series of optimistic assessments of new, large-scale flows of foreign capital, the slack in unused capacity, favorable commodity prices and high demand, and expansion of domestic consumption based on rising income. Even accepting these dubious projections, 2004 will barely recover a part of the losses in living standard suffered in 2003. The prospects for substantial growth in 2004 even under optimal external conditions (rising commodity prices, expanding markets, new trading agreements) are dubious. This is particularly the case in relation to the domestic market. The proposed multi-year budget allocates billions to meet interest payments. The substantial cuts in domestic public spending impose a serious constraint on domestic growth, while the enormous outlays to foreign creditors will not be compensated by any large-scale, long-term investments.

In other words, 2004 will see at best a very weak recovery (least of all on a per capita basis) and greater inequality, intensified plunder of the environment, and continued human rights violations. This will result in mass disenchantment with the broken promises of the

Da Silva regime. Even worse, 2004 will see the increased presence of the rightwing parties in the government along with long-term financial agreements that prejudice any strategic alternative development strategy.

The consolidation of the richest and most powerful and retrograde economic elites as central economic actors, in terms of public financing, private profiteering, and governmental policymakers, means that Brazil is in for a period of socially regressive development, based on an extraordinarily precarious set of external circumstances. Volatility and high risk accompany the Da Silva economic team's high dependence on rising commodity prices, flows of speculative capital, expanding external markets, and continued compression of the income of workers, farmers, and public employees. Commodity prices historically have gone through predictable cycles of high prices, world-wide expansion of production and borrowing on unpredictable future returns leading to over-production followed by a sharp decline in prices and demand, resulting in sharp reductions in regime revenues, highly indebted producers, and severe deficits and balance of payment problems. This in turn leads to the flight of speculative capital, precisely when the regime seeks investment to compensate for external imbalances, deepening the financial crises, and putting enormous stress on the entire financial system. The attempt by the regime to impose greater austerity on the population which is already being squeezed to finance creditors and to increase profits for investors ('making them competitive') is likely to promote widespread social unrest and extraparliamentary activity. Given the transfer of almost all strategic financial, mining, trading, and manufacturing industries to foreign capital, and given the 'autonomy of the Central Bank' (closer linkage to overseas bankers), the regime lacks the necessary economic levers and resources to intervene in the economy and stimulate growth.

The systematic shift in the governing PT's social base, the disenchantment of public employees, rank-and-file industrial workers, and rural landless workers, and the large-scale recruitment of new members on the bases of small favors, patronage, and party appointed jobs, means that the regime does not have a reliable social base to sustain it in times of economic crises. The 'neo-Lulistas' are easily recruited with promises of jobs and leave quickly when budgets are slashed and job promises fail to materialize, emptying the party-state apparatus of its electoral activists. To sustain the orthodox neoliberal policies and the regime's alliance with the right requires more than

3.5 per cent growth in 2004 and a period of sustained expansion of the world economy, a decline in security threats, the ending of protectionism in the US and Europe, and the total immobilization of labor and peasant organizations. All these assumptions are highly improbable.

The US and the EU are highly unlikely to end farm subsidies and protection. The economic recoveries of the US and the EU are very problematic and even less so beyond 2004. The US will continue its military colonial policies, leading to permanent global insecurity, and it is likely that the social movements, trade unions, and leftwing parties will resist the regime's efforts to neutralize them.

By the end of 2003, it was clear that the Brazilian left had suffered severe political and social defeats. But it is also true that a substantial sector of the left is now wiser, and conscious of the fact that Lula is a formidable adversary not a friendly ally.

Lula's rightward turn has spurred a range of explanations. In the first few months of his regime, Lula loyalists argued that the orthodox neoliberal policies were 'tactical moves' to stabilize the economy before turning to social reform. This argument lost credibility as Lula's policies, appointment alliances, and legislation all converged into a logical coherent orthodox neoliberal strategy. Subsequently, a variety of other explanations emerged. In the opening section of this essay, we pointed to a multiple factor explanation that encompassed:

(1) the evolution of the PT from a movement-based party into an electoral machine built around Lula's persona and his personal coterie of advisors;
(2) the rightwing alliances and elite financing of many of the governors, mayors and other elected PT officials leading up to the presidential elections of 2002;
(3) the changing class composition of the PT Party Congress, highlighted by the predominance of middle-class professionals, party officials, and trade union bureaucrats, rising to 75 per cent in the last party congress; and
(4) the programmatic shift from a socialist agenda in the 1980s to a social welfare program in the 1990s, social-liberalism prior to the presidential election, and finally the orthodox 'Taliban' neoliberal practice of the PT presidency.

An additional argument by the Brazilian sociologist and former founder of the PT, Francisco de Oliveira, is the transformation of the trade union bureaucrats into a 'new class of managers of millions of dollars, directors of public firms and public funds' (*Critica Social*, November 3, 2003). Working closely with bankers and investors this 'new class' is now in cabinet positions or serve as Lula's advisors and share the neoliberal policies of the bankers and corporate directors who formulate Lula's economic strategy (*Folha de São Paulo*, October 29, 2003, A11).

These ex-trade union bureaucrats-turned-congressmen, cabinet ministers, and fund managers have strong ties with many of the existing CUT leaders, including its current head, Luiz Marinho. Their goal is to subordinate the workers to the regime, supporting greater restrictions on labor unions, undermining salaries and pensions, and above all preventing any unified mass-based direct action against Lula's neoliberal policies (Oliveira, *Critica Social*, November 3, 2003). The bureaucratization and degeneration of the CUT over the past decade was noted early in the 1990s by the Brazilian sociologist Ricardo Atunes. Having collaborated with capital and previous neoliberal regimes prior to Lula, the trade union bureaucrats took advantage of their longstanding ties to Lula to enter his regime and openly promote the interests of capital against labor. The revival of the Brazilian left thus faces formidable new barriers in pursuit of jobs, land, dignity, and justice—their former political leaders and trade union officials, now allied with the US, the IMF, and the Brazilian elite, and backed by the resources of the state and the support of the mass media.

The social opposition to Lula has so far been confined to sectoral protests and strikes by public employees, metal workers, and urban squatters—with mixed results: the public employees were not able to overturn Lula's slashing of pensions, while the metal workers were able to secure some improvements in salary. The rural landless workers' movement, including the MST, has continued land occupations, but under increasingly repressive conditions and with incomprehensible illusions among some national leaders about the nature of the regime. The progressive church, the CPT, CNBB, and even Caritas have voiced strong criticism of Lula's orthodox neoliberal priorities, but like the MST hope the regime will change in 2004. The CUT has demonstrated neither the will nor the capacity to mobilize against Lula, divided as it is between a collaborationist leadership and an increasingly discontented base. The four expelled PT parliamentarians have made a courageous break with the PT and are engaged in numerous mass meetings to build a new party. It remains to be seen how effective they are in regrouping and unifying the disparate but growing mass opposition to Lula.

Most of the disenchanted workers are withdrawing support from Lula rather than joining new parties. This can change as more and more of the populace sees through Da Silva's populist theatrical 'plain talking' and understand his servile and unconditional support for foreign investors, agro-exporters, and speculators. In this regard, Lula's 'style of politics' is a central problem that requires serious analysis and critique since it plays a big role in mystifying the poor, even as he strikes blows against their living conditions, social demands, and hopes.

Da Silva had mastered the art of combining *symbolic gestures* to the movements and common people with *substantial* economic concessions and resources for the rich, including the foreign rich. For the poor, he enacts emotional scenarios accompanied with acts of personal compassion. He cries real tears faced with child poverty, then abruptly follows with a major reduction in social spending and massive transfers of wealth to the creditors. He meets with the MST and playfully puts on one of the organization's hats and then in a press conference ridicules their agrarian reform program, reassuring the big agro-exporters with increased subsidies. Lula has mastered the pseudo-populist demagogy of the US's ex-President Clinton by telling the poor he 'feels their pain,' while he proceeds to push one regressive measure after another, lowering the minimum wage, facilitating the firing of workers, and criminalizing the social movements. We can sum up Lula's style of politics as 'populist in form and reactionary in content.' Over time content will determine form, just as material existence will influence consciousness, but it is not an automatic process. The year 2004 will not lead to a collapse of Lula's regime but it will be a turning point for the social movements, leftwing parties, and the Church: They can begin 'the long march' toward the construction of a new mass political party movement, where the direct needs of the people orient the organization and direct action becomes the main vehicle for struggle.

LULA YEAR 2: DEEPENING NEOLIBERALISM

Despite the disastrous socioeconomic results of the first year of Lula's implementation of his neoliberal agenda, he resolved to continue, extend, and deepen these policies, in both domestic and foreign affairs. Lula's foreign policy is an extension of his domestic policy, and his domestic policy is an extension of his foreign policy. At the beginning of 2004 he announced plans to privatize Brazil's infrastructure, deepen Brazil's role as a commodity exporter, increase Brazil's dependence on overseas markets, continue domestic austerity and high interest rates to accommodate foreign and domestic bankers and speculators, approve a modified version of ALCA and deepen Brazil's role as a raw material exporter to China. In pursuit of these neocolonial policies, the regime embarked on a foreign policy which included sending 1,500 Brazilian military forces to Haiti to protect the US puppet regime put in power by the US Marines; the mobilization of 15,000 troops to the Colombian border in coordination with the terrorist Uribe regime, minimal diplomatic and political relation with the Chavez government in Venezuela and Castro in Cuba, and an 'open door' investment policy to financial speculators in the US and Europe. In addition, Lula increased ties with China based on imports of manufactured goods and exports of minerals and agricultural commodities.

Lula's economic policies, 2004

Lula reaffirmed the continuation of his neoliberal policies, support for the IMF, and his unconditional backing for his neoliberal economic team throughout 2004. He clearly and loudly rejected Joao Pedro Stedile's and the so-called 'left' PT's call for a 'turn' to the left or change in his cabinet.

In the first quarter of 2004 industrial production fell by 1.2 per cent compared to the last quarter of 2003 despite more working days in January–March 2004 than in October–December 2003. Despite record low international interests rates, Brazil has the world's highest rates (over 16 per cent) and threatens to go higher to attract speculative capital and cut off any possible recovery. While year-to-year industrial growth was positive, the sectors that showed the greatest increase were the foreign-owned utilities (electricity, telecommunications), which successfully pressured the Lula government to raise rates, thereby increasing profits to the multinationals and forcing up the costs to consumers. Electricity saw a huge rise in net income of 94 per cent, and telecommunications of 53.2 per cent, during the first four months of 2004 over the same period in 2003, despite the decline in the number of telephone lines in service because of growing unemployment. Other sectors such as steel, chemicals, and papers and pulp showed significant declines in net income even as raw material exports such as iron ore and soya continued to grow—until the sharp decline in exports from April (*Financial Times*, May 12, 2004, pp. 6 and 21).

In large part Brazil's continued stagnation, despite the boom in agro-mineral exports, was a result of the freeze in public investments as Celso Furtado noted in a speech delivered on May 4, 2004. At the same time, the regime earned the praise of the IMF as they renewed their agreement to continue following the restrictive monetary policies and open markets which allowed overseas speculators and bondholders 'to receive the highest returns of any emerging markets in 2003' (*Financial Times*, March 29, 2004, p. 11). The prognoses for the rest of 2004 are not favorable: export earnings which rose 12 per cent in 2003 may fall as the prices of Brazil's most important exports (soya, orange juice, and iron ore) decline due to the slowdown in China's economy and a 30 per cent decline in prices. Global interest rates are rising in the US, Asia, and Europe, putting pressure on the inflow of capital, and the volatility of the oil markets and the energy crises are likely to lower world demand and result in Brazil tightening even more the austerity programs in order to achieve the budget surplus targets of the IMF. As Cesar Benjamin pointed out, 2003 was not (as Lula apologists argue) a 'crisis' year, it was an optimal year for growth, and because of Lula's neocolonial and neoliberal policies Brazil failed to take advantage. Given the decline of opportunities for export growth and the continued weakness of the internal market, Lula can be expected to deepen the concessions to capital, cheapen the cost of labor, provide more tax concessions to investors, further deregulate the exploitation of the environment, and negotiate an agreement with the US on ALCA.

The 'lesson' that Lula draws from the economic disasters of 2003 is to move further to the right—to offer Brazil to the highest bidder not only in the US, but Europe, China, and to whoever is interested. In March, even the PMDB, Lula's rightwing coalition partner, attacked the government's extreme neoliberal restrictive monetary policy, lack of public investment, and concentration of credit to big business exporters at the expense of local small business people.

In order to pay for Lula's mass media electoral and publicity campaigns, Jose Dirceu, key advisor to the government, was exposed as heavily involved in bribe-taking from criminal syndicates. In attempting to put a 'moral veneer' on his regime, Lula outlawed bingo throughout the country, putting thousands of workers out of work and prohibiting one of the few recreational outlets for low-income workers.

Despite opposition from the entire spectrum of political parties from the Liberal Party, PSDB, PMDB, PTB, PCB, PS, the neoliberals running the regime's economic policy have Da Silva's unconditional support: Paolocci, Dirceu, Meirelles (head of the Central Bank) are a united team, united in present and future orthodox neoliberal policy. This 'government class unity' faced continued opposition calls by the 'left' PT ('Left Articulation') and some leaders of the MST to replace Paolocci and Meirelles. These 'appeals' have fallen on deaf ears because Lula is the self-proclaimed and forceful defender of neoliberalism and that is why he backs Paolucci and Meirelles. Even the rightwing São Paulo State Federation of Industrialists criticized the regime's pro-IMF extremist 'tight money policy' and high interest rates, pointing out that investors prefer to bank their funds rather than invest or spend them at a time of high unemployment and low levels of consumption (*Latin American Economy and Business*, March 2004, p. 11).

While inflation is declining, unemployment continues to grow: from 10.9 per cent in December 2003 to 12 per cent in February 2004. Throughout 2004 Lula traveled overseas with a huge entourage of big business executives looking for export opportunities and potential foreign investors to take over Brazilian assets. Data from 2003 demonstrate that the seven biggest privatized firms made record profits of more than R14 billion (just under $5 billion) in profits, much of which was remitted overseas instead of being re-invested in the country. High interest rates have been a lucrative source of super-profits for the foreign and domestic banking elites; high utility rates have increased profits for the gas, electrical, and telecommunication industries. The rate of profit for the banking sector was 21 per cent while for the non-financial sector it was 15.6 per cent.

It is clear that the privatized banks and their new owners are the main beneficiaries of the regime. Moreover, they provide the 'class base' for Lula's single-minded pursuit in 2004 of new foreign investors, more privatizations, and ongoing lowering of trade barriers. The results are an increase in overseas remittances, more dependence on volatile primary commodities, and greater vulnerability to speculative investments. The problem is that as commodity prices declined in mid-2004 and US interest rates rose, speculative capital was disinclined to invest so that the Brazilian 'recovery' will not take place. Moreover, with high debt payments there are few domestic and state resources to 'spark' a recovery. Brazil under the neoliberal policies practiced by Lula and his predecessor Cardoso has declined sharply in comparative terms over the past six years, from the 8th to the 15th biggest economy in the world. Between 1998 and 2004 the height of Cardoso–Lula neoliberal policy, Brazil's average annual GDP was 1.4 per cent, ranking 164 among 178 countries (*Global Invest*, May 2004).

In April, the regime announced a $5.1 billion investment and financing plan for big business. The plan—all supply-side policies— calls for tax breaks for industrialists who purchase new machinery, new state subsidies for research and development by private companies, state-financed international marketing campaigns, greater deregulation for business, including sharp cuts in government health and safety programs, environmental deregulation, and control over imports. The regime also proposed additional funding for some pay increases for civil servants and land settlements for landless peasants—largely as a result of a massive land occupation movement. Nevertheless, the implementation of this minimum program is highly unlikely to make much of an impact, since Finance Minister Paolocci assured bankers that the government would maintain its primary budget surplus of 4.25 per cent of GDP as well as neoliberal orthodoxy for the remainder of Lula's presidency (*Financial Times*, 1 April 2004, p. 2).

The social crisis deepens

In the first quarter of 2004 Brazil had a trade surplus of $6.170 billion according to the Secretary of Foreign Commerce. This largely went into the pockets of the agro-mineral elite and the overseas and local financial holders of Brazilian debt.

In contrast to the generous payments ($30 billion) to the foreign creditors during the first half of 2004, and the record super-profits for the agro-mineral exporters, Lula proposed a 1 per cent real increase in the minimum monthly wage from 240 to 260 Reales (3 Reales = $1), approximately $86 a month, one of the lowest in South America. Lula's argument that the government could not 'afford' a higher raise was ridiculed by all the opposition parties and even by some members of his own party. Even the conservative parties proposed an increase to 275 Reales per month. Lula's minimum wage policy makes a mockery of his promise to 'double the minimum wage in four years.' In fact, even with low-level inflation, the minimum wage will at best remain unchanged throughout Lula's presidency, condemning over 4.5 million workers to near indigency.

Lula's promotion of the highly mechanized, capital-intensive 'agro-mineral export strategy,' the low level of public investment, and the decline in living standards has weakened the domestic market and exacerbated unemployment. May 2004 registered the highest recorded unemployment figures—13.1 per cent according to data from the Brazilian Institute of Statistics and Geography *(Instituto Brasileño de Geografia y Estadística*—IBGE). Worse still, the unemployment rate is worsened between 2003 and 2004. The IBGE also indicated that real average income in April 2004 was 868.50 Reales (US$280), a fall of 3.5 per cent with respect to April 2003. The unemployment rate in metropolitan São Paulo has remained in the 20 per cent range. An indication of the gravity of the unemployment crisis in São Paulo is found in the subway systems announcement of 30 job vacancies, to which 126,000 applied—more than 4,000 applicants for each position.

In a major speech during his visit to China, Lula called on the United Nations and the G-7 countries to join in a crusade against hunger. Lula's speech was typical of his 'public relations' rhetoric to bolster his international image as a spokesman for the poor of the Third World. The reality of growing hunger and food insecurity under his regime belies his speeches in international forums. Falling living standards, the concentration of land ownership, pension cuts, high unemployment rates, the increase of low-paid, 'informal' employment, the rising food costs increasingly based on food imports, have all reduced food consumption and food security. Even if we accept the very questionable government estimates of 53 million Brazilian suffering 'food insecurity'—malnutrition—Lula's *Plan Zero* had not provided food to even one sixth of the population over the past year. The *Plan* is a total failure from a *structural* point of view, as it has not changed land tenure or the productive and distributive systems that create hunger. Lula's zero hunger program continues to be at best a 'charity program' to control potential poor voters. Rather than eliminate poverty, the zero hunger program has enriched its administrators, fostered corruption, and reinforced traditional patron–client relations.

Given the large trade surplus of R20 billion and the surplus of R41 billion (a little less than $5 billion) in the social security fund for the first quarter of 2004, it is clear that there are ample funds available to increase the minimum wage substantially, to expand public investment, to create jobs, and to finance land distribution. The real problem is not 'lack of funds,' as Lula argues, but the political—economic strategy, the class interests, which the regime has embraced. Under Lula, the surplus is distributed to different sectors of the ruling class: the foreign and local financial interests, agro-mineral exporters, and the giant, foreign-owned manufacturing and petrochemical companies.

In 2002 labor received 50 per cent of national income; in 2003 this declined to 36 per cent as over two-thirds of income was further concentrated in profits, interest, and rents reaped by the capitalist class. This 'development' also relates to the development of poverty in the country. The real cause of increased poverty is found in the greater concentration of wealth, which results from the 'carnal relations' of the regime and the capitalist class.

To maintain the dominance of capital, Lula proposes new legislation to weaken the capacity of workers to call strikes by making arbitration between employers and workers all but obligatory (*Financial Times*, April 16, 2004, p. 5). The anti-labor legislation includes greater 'flexibility' for the capitalists in firing and hiring workers, employment on short-term contracts, lowering severance pay, and other measures designed to increase the power and profits of business. The CUT and *Forza Sindica*, the two major trade union confederation, are in agreement, because they will be given greater authority to 'manage' union pension funds, and increased centralized control over unions, though they will lose the automatic deduction of union dues.

As Lula's neoliberal policies deepen poverty, extend unemployment, and increase the numbers working in the 'informal economy' to nearly 50 per cent of the labor force, so too does crime increase to record levels. Drug trafficking provides a livelihood directly or indirectly for millions of poor Brazilians, who are unable to live on a minimum wage of $80 a month.

Violence between gangs, large-scale assaults on beaches, and street crime are endemic. There are two approaches to crime: the conservative policy is to increase police repression, militarize the low-income neighborhoods, and longer prison sentences for offenders; the progressive approach is to invest and create jobs and housing, raise incomes, and provide access to livable wages as an alternative to crime.

Lula's regime has increased police powers and in May 2004 militarized the major *favelas* of Rio de Janeiro (*Financial Times*, May 5, 2004 p. 2). The prisons are vastly overcrowded and riots are frequent as repression increases the rate of incarceration but fails to lower the crime rate.

Lula's policy toward 'alternative solutions' to unemployment is also totally inadequate. His regime proposed creating jobs via subsidies to employers (First Job Program), but in the first six months of 2004 only 725 jobs were created (*Financial Times*, May 5, 2004, p. 2). Secondly, Lula plans to 'conscript' 30,000 extra young men into the army in an agreement between the Employment and Defense Ministry and business associations. Without adequate funds or experience the military cannot fund and train the new recruits for civilian jobs. Given the Lula regime's commitment to meet debt obligations, there are no funds available for large-scale public works projects that could make an impact on the growing millions of unemployed young people. Ultimately, Lula's crime policy has only one direction—more repression, more violence, and greater citizen insecurity.

Opulence in the midst of misery

In mid-June 2004, President Da Silva met with his close ally, multi-billionaire Fernando de Arruda, and 15 other big capitalists to plan the state's economic priorities. A few days later, Fernando celebrated his birthday with 8,500 guests, 300 of whom flew in their private airplanes to his *estancia*. In late June, Lula rejected attempts to raise the minimum wage from Reales 240 to 275 (US$77 to $88) a month. He defended a real increase of 1.5 per cent to $83, arguing that the extra $5 a month for the impoverished, hungry working poor was against his fiscal austerity policy.

Nothing symbolizes Lula's obscene servility to the festive super-rich like Fernando de Arruda who can squander millions of Reales for a birthday party and the president's arrogant contempt for the poor struggling to survive on $2 a day. Every month of Lula's regime provides more evidence that his policies are widening inequalities: in 2003–4 the number of millionaires grew from 75,000 to 85,000 while the vast majority of workers saw their living standards decline by 12 per cent. To further the pillage of Brazil's national resources, Lula is preparing on August 15 to auction off to foreign capital petroleum regions with known reserves of 6.6 billion barrels of petroleum (as determined by the state petroleum company, Petrobras), amounting to 50 per cent of the proven reserves in what will surely be remembered by historians as the 'Great Petroleum Robbery.' Cesar Benjamin asks: 'What adjectives does a government merit which acts like this?' We can suggest many: 'sell out,' *entreguista*, but even the worst epithet cannot match this unconscionable act of betrayal of many future generations of Brazilians.

Land reform: Lula and the MST

The greatest failure of the Da Silva regime is in the area of agrarian reform. Despite repeated promises to the landless rural workers, the Lula regime has expropriated far fewer farms than his neoliberal predecessor Fernando Cardoso. Between January and April 2004, the regime provided land to only 7,000 families against a promise to settle 115,000 families for the year. Altogether, for the first 16 months, Lula had settled 21,000 landless families against a promise of 230,000. As a result of Lula's false promises and unwillingness to advance the cause of 24 million of the most exploited rural workers, the MST launched a national land occupation campaign on March 27. By the end of April, 33,411 families had occupied over 135 *latifundios* in 20 states. Pernambuco, one of the poorest states, witnessed 32 land occupations involving over 8,000 families (data from MST reports). By the middle of May, another 14 *haciendas* were occupied in Pernambuco. By the middle of 2004, there were tens of thousands of families—some estimates say over 200,000 families—occupying land in the most primitive conditions, waiting for INCRA to expropriate the land and settle the families. INCRA follows Da Silva's directives 'to follow the law' and has refused to expedite the expropriation process. The reactionary judicial system has created innumerable legal delays as the landlords successfully block the process through continual appeals.

Lula has failed to bring about agrarian reform because it is incompatible with his agro-export strategy, and his links to plantation owners and agro-business multinational corporations. After 18 months it is clear that there are five basic structural contradictions between the needs of the landless workers and the regime:

(1) between the policy of concentrating and centralizing production in the hands of the agro-export elite and the demands of the landless for land redistribution and expansion of the domestic market;

(2) between concentrating financial and commercial credit, transport and technical assistance to agrobusiness and their denial to land reform beneficiaries;

(3) between privileging and legitimizing agro-business expansion and criminalizing social action and land occupation by the landless rural workers;

(4) between the regime's massive transfers of wealth to foreign and domestic bankers and the drastic reductions in social services and

public investments for small-scale producers, rural proletarians and Indian and Black communities; and

(5)   between the subordination of Brazil to the imperialist division of labor and the effort of landless workers, cooperatives, and small farmers to deepen and extend the linkages to the domestic market.

Lula has deeply exacerbated the *polarization* within the Brazilian class structure and economy—between exporters and local producers, between financiers and manufacturers, between private owners of the means of production and the public employees. This polarization is most sharply evident in the countryside where agro-exporters receive 90 per cent of the credit and major incentives to export and grow by 20 per cent a year, while small farmers and cooperatives producing food for the domestic market are on the verge of bankruptcy and earnings are stagnating. Starved of resources, facing competition from subsidized food imports, the cooperatives resulting from the land reform are facing a disastrous future.

The problems for the agrarian movements—including the MST—are not merely the lack of progress in agrarian reform, but the agro-export-centered policies, which threaten to *reverse* the progress of the past 20 years. The regime brings together the most formidable configuration of agro-business power in recent years, and they have his unconditional backing. Lula's support for ALCA ('lite'), and his signing of new trade and investment agreements with China are all indicative of a powerful strategic alliance with the agrarian classes most hostile to agrarian reform and small producers for the local market.

The problem of the agrarian movements (landless and small farmers) is eminently *political:* Lula's regime represents agro-business, and the popular rural classes have no political representatives or leaders. The strategy of the MST giving 'critical support to the Lula regime' (Joao Petro Stedile: interview, Paris, September 5, 2004) has boomeranged. MST support of Lula has strengthened Lula in alliance with the agro-business elite while weakening the opposition social and political movements.

For nearly 15 months the agrarian movements stagnated, tens of thousands of landless workers were 'camped', waiting for Lula to respond to the 'pressure' of the 'left' PT and other 'friends of the MST' in the regime (Betto, Rossetti). This is a time of suffering, waiting, disillusionment, and finally frustration, leading to a series of localized land occupation independent of the central leadership of the MST. The reliance of the MST national leadership on its 'close ties' with Lula to secure positive changes was a tragic illusion with very negative results for all concerned. All around the MST there is growing rejection of Lula's embrace of the ruling class, the IMF and foreign banks. Less than a third of the population were supporting Lula's economic policy by June 2003. The public employees organized strikes and 50,000 marched to Brazilia; thousands of long-term members of the PT abandoned the party. Dissident PT Congress members expelled from the party formed a new party in May 2004: The Party of Socialism and Liberty. The PSTU organized the 'Coordinator of Popular Struggles.' The MST did not participate in any of these new movements, whose policies they share. The national leadership remained tied to the impotent, opportunist 'Left Articulation' fraction of the PT, unable to influence the Lula regime, but certainly weakening the growth of a real progressive political movement capable of offering an alternative to the ultra-liberal policies of the Da Silva regime.

It is difficult to imagine how the MST leaders, after 18 months of savage attacks on workers' salaries, pensions, labor conditions, minimum wages, and land reform, can justify 'critical support' of the most reactionary financial agribusiness regime in recent Brazilian history. In this connection, the MST's massive land occupation movement in April/May 2004 is a major step forward, a recognition that nothing progressive will come from Lula's regime. However, even as the land occupation occurred, Lula completely ignored them: he was off to China with 400 bankers, agro-businessmen, and industrialists signing agreements to expand their *markets*, their *profits*, and offer lucrative deals to their Chinese counterparts. Lula plans to invest tens of billions of Reales for joint ventures with China to promote agro-mineral exports. He has yet to provide any meaningful funding for the uncultivated lands occupied by the MST activists.

The conclusion is clear: the political problems facing the MST cannot be solved simply by social struggles—albeit that is far better than waiting for change via an elected neoliberal politician. *Social* movements like the MST in Brazil cannot rely on electoral parties, particularly those so deeply embedded in the capitalist institutional establishment like the PT. Electoral politics has corrupted and co-opted every left party over the last 30 years—and the PT of Brazil is no exception. The political choice is not between becoming a partner or ally of an electoral party or remaining merely a social movement, but to move from sectoral or national social protests to a strategy for building a political party for taking state power. Lacking their own party and dependent on the electoral strategy of the PT, the MST have turned to what they do best—organizing social struggles (land occupations) and social protest. However, facing a coherent, determined regime bent on transforming Brazil into a neoliberal 'partner' of Chinese elites and an appendage of US and European imperialism the old militant tactics of 'organizing the masses and pressuring the state' is not likely to have much effect on the global policies of the Lula regime.

Conclusion

The transformation of the politics of the Workers Party (PT) from a 'party of the masses' into a 'party of big business' is found in the long-term, large-scale structural changes *within* the party and its relationship to the state. The decisive shift was from mass popular social struggles to electoral politics. As a result the PT became an 'institutional party,' embedded in all levels of the capitalist state and attracting a large number of petit-bourgeois professionals (lawyers, professors, journalists), trade union bureaucrats, upwardly-mobile ex-guerrilla, ex-revolutionaries recycled into the electoral arena. The electoral party leaders—congressional deputies, senators, governors, mayors, councillors—and the 'political machinery' of electoral politics, took over the key posts in the party, defined its 'operational program' (as opposed to the 'rhetorical program') and established the tactics and strategy. A process of substitutionism took place, where the electoral apparatus replaced the popular assemblies, the elected officials displaced the leaders of the social movements, the institutional maneuvers of the national political leaders in Congress substituted for the direct action of the trade union and social movements. And the leadership cult of Lula and his clique became the embodiment of the masses and replaced the popular forces acting on their own behalf.

The historical and empirical data demonstrate that the elitist electoral leaders *embedded* in the institutional structures of the capitalist state ended up competing with the other bourgeois parties over who could best administer the interest of the foreign and domestic, agrarian and financial elites.

The fundamental change in the class composition of the PT was based on the shift to electoral and institutional politics: it became the party of ambitious, *upwardly-mobile* lower-middle-class professionals whose *social reference* was the capital class.

The 'new class' of PT electoral politicians looked *upward* and to their *future* ruling class colleagues, not *downward* and to their former working-class comrades.

The change in *class consciousness* reflected the change in *material conditions* of the elected politicians. For the PT politicians mass voting support among the working-class, landless workers and urban *favelados* became bargaining chips to negotiate favors with big business.

To demonstrate that the PT was an acceptable interlocutor with Brazilian big business, it dumped its Marxist and socialist identity early on (even though inconsequential minority factions continued to call themselves 'Marxists'). As the PT accumulated allies among big business it moved to attract the support of foreign capital and certification from Washington—signing agreements with the IMF, declaring in favor of privatization, setting aside record sums to 'honor' foreign debt payments. In the period of global imperial politics, 'capitalist logic' embedded in the institutional politics of the PT led inexorably to total subordination to imperialist division of labor under the leadership of direct representatives of local and foreign big business and finance—as evidenced in Lula's key economic ministers.

PT's evolution toward the logic of neoliberalism was led by Lula, who in turn was influenced by the institutional party. The PT's transformation into a party of international capital was accompanied by the transformation of the major trade union confederation, the CUT, from an independent class-based union into an appendage of the Ministry of Labor. The CUT followed a similar process of 'state institutionalization' and 'substitutionism' as the national leaders pre-empted the factory assemblies in making decisions and relocating activists from the streets to the offices of the Ministry of Labor. The upwardly-mobile trade union officials looked to becoming congressional candidates and ministers, or administrating pension funds rather than to organizing the unemployed, the urban poor in general strikes with the employed workers. Thus the parallel transformation of the PT and CUT avoided any rupture between them.

The key theoretical point is that the bourgeoisification of the PT and its leaders was not the 'inevitable consequence of globalization.' It was the result of the changing class ideology, composition, and strategy of the political party—changes which led to institutional assimilation and ultimate subordination to the dominant sectors of the ruling class—Brazilian and Euro-American. Theoretically, this points to the profound limitation of electoral-institutional politics as a vehicle for social transformation or even consequential reforms. Social transformation is far more likely to occur from the direct action of independent, class-based social political movements oriented toward transforming the institutional basis of bourgeois state power.

4

Social Movements and
State Power in Ecuador[1]

The electoral victory of Lucio Gutiérrez in 2003 was greeted by the same sense of optimism and wild expectation of a new direction and alternative politics that greeted the election to the presidency of Lula in Brazil and that has surrounded Hugo Chavez's declaration of the Bolivian Revolution. Also, developments in Bolivia (Evo Morales, the leader of an organization of coca-producing peasants, almost coming to power through electoral means, the overthrow of de La Lozada through a popular mobilization) awakened hopes on the left of a new dawn in Latin American politics. Even the ascendancy of Nestor Kirchner to state power in Argentina awakened the same hopes and expectations for a fundamental change of direction in national policy—at least as regards the IMF and a policy of non-payment of interests on the foreign debt in the interest of productive investment in Argentina's economy. In each case, and collectively, these political developments have been widely viewed on the left as growing evidence of the demise of neoliberalism and the power of the US to shape economic policy in the region, and part of a new wave of progressive regimes oriented toward, and committed to, an alternative popular form of national development, constituting thereby an anti-US axis in foreign policy—Argentina, Brazil, Ecuador, and Venezuela. Notwithstanding the title of a recent Italian film (*How Bush Won the Elections*), the electoral victory of Gutiérrez was also widely interpreted on the left as a setback to the efforts of the US administration to dominate economic and political developments in the region—to reassert its hegemony and what even some of its advisors see as a project of imperial domination.

A year after Ecuador's presidential elections, and a year into the Gutiérrez regime, a number of questions have arisen. They include: (1) how realistic were these expectations, on the left, of progressive change in the region? and (2) What is the political significance of Gutiérrez's rise to power and his government? These questions have already been answered in regard to Ignacio Silva (Lula)'s ascendancy to state power in Brazil. In this case the left's expectations and illusions were soon shattered. And Gutiérrez? His election to state power was also widely viewed both within the country and outside as a victory for the left and a blow to the US and its agenda for Latin America.

This chapter places this question in the context of a broader set of theoretical issues that surround the relationship between the state and social movements. One of these issues has to do with the question of state power and the different paths toward achieving it—electoral politics, the path preferred by the 'political class' because it is predicated on limited political reforms to the existing system, or mass mobilization of the forces of resistance and opposition. This is the path taken by most social movements. It is oriented toward more fundamental or radical change in the existing system. In addition, there are two other conceptions of political practice and power. One is associated with a postmodernist perspective on a new form of 'politics' and the emergence in theory (that is, academic discourse) of the 'new social movements' (Slater, 1985; Escobar and Alvarez, 1992; Melucci, 1992; Burbach, 1994; Calderón, 1995; Helman, 1995; Hunter, 1995).

In this perspective, the way to bring about social change is not through political action in the struggle for state power but through social action involving the construction of an 'anti-' or 'no-power'—in social relations of coexistence, solidarity, and collective action (Holloway, 2001, 2002; Negri, 2001). Another approach to social change is associated with the NGOs involved in the process of international development. The ostensible aim and central objective of this strategic 'project' are to partner with governments and organizations of overseas development assistance in promoting an improvement in the lives of the poor—to bring about conditions that will sustain their livelihoods and alleviate their poverty—and to do so not through a change in the structure of economic and political power but through an empowerment of the poor. This approach seeks to build on the social capital of the poor themselves, seeking thereby to bring about improvements in their lives within the local spaces available within the power structure. It is predicated on partnership with likeminded organizations in a shared project (the alleviation of poverty, sustainable livelihoods). Rather than directly confronting this structure in an effort to change the existing distribution of power, the aim, in effect, is to empower the poor without disempowering the rich.

Under conditions available across Latin America, and experienced in different ways in virtually every country, each of these modalities of social change and associated conceptions of power can be identified or have their adherents. However, in the specific conjuncture of conditions found in Ecuador the postmodernist perspective is irrelevant. All of the social movements in the country, from the most to the least consequential and dynamic of these movements, are involved in a struggle for state power. As for the political dynamics of the struggle the issues are diverse, but the most critical turn on the relationship of the social movements to the state. This is the central object of our analysis.

THE CONFEDERATION OF INDIGENOUS NATIONALITIES OF ECUADOR: THE FORMATION AND DYNAMICS OF A POPULAR MOVEMENT

The history of Ecuador's indigenous people,[2] as told by the ECUARUNARI, a social and political organization of the highland Quichua, is a history of class struggle against 500 years of 'exploitation and oppression' and expropriation of their land—this as recently as the 1960s and 1970s with a system of agrarian reforms designed in support of the big land proprietors, the *latifundistas*, to ensure more productive use of their large landholdings and estates, while the relatively poor and unproductive part of their holdings were distributed to the mass of landless and near landless 'indigenous peasants,' generating in the process the *minifundio* and widespread poverty as well as a large-scale exodus and migration to the cities, transforming the country's indigenous peasants into a proletariat (ECUARUNARI, 1998: 46–7).[3] Another part of this history is a tradition of struggle and uprisings, and, in the twentieth century, a turn toward class-based organizations such as the FEI (*Federación Ecuatoriano de los Indios*) and FENOC (*Confederación de Organizaciones Clasistas*)[4] as means of advancing the struggle for social justice, ancestral collective rights to the land ('mother earth'), improved access to the means of production, and the productive resources needed to sustain their livelihoods and communities, territorial control—and, above all, freedom from diverse forms of exploitation and oppression *('los tributes . . . los diezmos . . . los castigos, los azotes . . . la opresión y el despotismo')* (ECUARUANARI, 1998: 21).

The 1980s presented a new conjuncture in the history of this struggle and in the formation of a popular movement. Implementation of a new economic model in the form of structural adjustment policy reforms created conditions that in the late 1980s reached crisis proportions in both the urban centers and the countryside.

These conditions included a precipitous decline in the level of manufacturing employment, a fall of 50 per cent (in the second half of the decade) in the construction industry, of critical importance for the growing mass of rural migrants, and a dramatic fall in the purchasing power of wages—29 per cent from 1980 to 1985 and a further 8 per cent annually from 1986 to 1990 (Zamosc, 1993: 289; ECUARUNARI, 1998). In addition, those indigenous communities and peasant farmers that retained access to some land were hit by a decline in demand for their product, increasing costs of inputs (fertilizers, etc.), a reduction in (and increased costs of) available credit, runaway inflation in the prices of consumer goods, and a general cutback in government services (Chiriboga, 1986; Zamosc, 1993).

Under these and other such conditions, in addition to concerted efforts of the government to contain the level of violent conflict, the rhythm of class struggle waged by indigenous peasants declined, at least relative to the level of struggle by public sector workers who maintained their rhythm of struggle over the course of the decade— 33 per cent of all confrontations and protest actions vis-à-vis the state versus 3.2 per cent for the indigenous communities and peasant farmers. But Sánchez Parga (1993: 55) points to a trend toward the persistence of class struggle waged by the indigenous communities and peasant farmers, and a high level of social conflict and political confrontation from 1980 to 1984 and again from 1988 to 1992. In the first period, observers identified 59 protest actions and confrontations involving 'peasants' (the land question, etc.) and 32 involving the indigenous struggle for ancestral rights, territorial autonomy, and social justice. From 1984 to 1988 there was a general decline in these struggles (only seven recorded protest actions) while the rhythm of social conflict associated with these struggles accelerated in 1988—20 cases of conflict involving the indigenous communities as 'peasants' over the next four years and seven involving the 'indigenous' question of 'ethnic identity.' The study by Sánchez Parga also points toward a shift in the dynamics of this struggle. From the perspective of ECUARUNARI (1998: 160) the cycle of struggle (up in the early 1980s, down from 1984 to 1988, and then up again) was related to, and explainable in terms of, the level of government repression, each increase provoking a higher level of political conflict.

Some political sociologists in this context wrote of the emergence of new social movements that appeared to escape the problematic of class struggle, raising the specter of postmodernism in the interpretation of these movements (Escobar and Alvarez, 1992). As it happened these 'new social movements,' characterized by a 'heterogeneous' social base and a concern with diverse non-class issues ranging from the defense of human rights, protection of the environment, democratic development, women's rights, etc., materialized in the relative lull in the class struggle from 1984 to 1988, allowing for this postmodernist interpretation. But the wave of these movements dissipated almost as soon as it appeared, persisting only in the 'theorizing' of sociologists ensconced in their offices and tuned into cyberspace.

In the wake of this new wave of social movements there emerged another based on a centuries-old class struggle for land and—in some contexts (primarily Bolivia, Ecuador, Mexico, and Paraguay)—an indigenous ethnic struggle for social justice and dignity, ancestral rights over mother earth, cultural and political identity, territorial autonomy, and a pluri-national state. The latter would become the major political project of a nationwide confederation of Ecuador's indigenous nationalities in the highlands, the coastal areas and Amazonia, formed in 1986.[5] The formation of the Confederation of Indigenous Nationalities of Ecuador (CONAIE) occurred in the context of a struggle with two major definable dimensions, one involving the question of the land and a class struggle of peasants against landed proprietors and the state; the other a struggle of indigenous peoples to 'construct their identity in a heterogeneous society' (Chiriboga, 1987).

It is possible to trace the history of CONAIE as a social movement over the next two decades in the form of a succession of political conjunctures. The first of these conjunctures was in 1988, two years after CONAIE's formation as a social movement and almost a decade before the formation of Pachakutik, a political movement organized to bring about 'a new country' based on a 'pluri-national and multi-ethnic state.' In this conjuncture, the main concern for CONAIE was to raise awareness within the movement about precisely 'who the enemy was,' namely 'the rich, a dominant class that has the power to dictate laws against another class, the poor, who have no power and suffer the consequences of these laws' ('*Antecedentes al surgimiento de Pachakutik,*' *Riccarshun*, December 1988; ECUARUNARI, 1998: 255).

The ultimate aim of CONAIE (and Pachakutik) was '*derribarles del poder del estado*' (to remove the 'oligarchs,' the 'enemy of the people' who have caused 'the poverty and misery of millions of Ecuadorians' *from state power*). The means for bringing about this awareness (what Marxists traditionally define as 'class consciousness') was a campaign to elect a popular constituent assembly to see through the political project of securing social justice for Ecuador's indigenous peoples within a new pluri-national state. To this end CONAIE's base were mobilized against government policies, particularly with reference to the underlying neoliberal model of free market capitalist development.[6] Each of CONAIE's constituent organizations 'actively participated' in this process (Pachakutik), 'consolidating the new [political] space with a proposal for an alliance with other social sectors of the popular movement, and seeking to overcome problems of internal disunity, particularly as regards the 'Amazon Question,' a matter of an urgent (and as it turns out) persisting political crisis (ECUARUNARI, 1998: 257).

The next critical conjuncture in the indigenous movement was in 1990 with (1) the government's implementation of a series of economic reforms and austerity measures designed in Washington (Williamson, 1990) and mandated by the IMF; (2) heightened class conflicts over land; and (3) a range of broad political demands, including the abandonment of the neoliberal economic model and the implementation of a series of alternative legal and constitutional reforms pressed upon the government in the form of direct actions initiated, on May 28, with the takeover of the Santo Domingo church in Quito but followed by similar actions in the highland provinces. These actions, it turned out, 'sparked the flame that lit the straw' of a conflagration that soon consumed the whole country (Lluco Tixe, 1998). With the supportive coordinated actions by diverse groups and organizations in the urban sector, especially the church-based communities in popular *barrios*,[7] the mobilization by CONAIE was transformed into an uprising of historic significance, an important reference point for a series of subsequent surges in the movement over the course of the decade, culminating in the January 21 2000 uprising. Major achievements of the 1990 uprising included the emergence of the indigenous movement as a new actor on the state of national politics and the unity of the majority of indigenous nationalities organized by CONAIE, particularly in terms of the 'contradiction' between the movement in the highlands and the Amazon, but also between the two discrete dimensions of the movement, namely the 'land question' and the struggle for cultural and political identity. In the heat of the uprising these two dimensions of a protracted struggle were combined into one, with the diverse sectors of the movement coming together (Lluco Tixe, 1998).[8]

The next major conjuncture in the movement, in 1993, occurred in the context of a referendum on the government's proposal to privatize the social security system and continuing class struggle over land.[9] As for the government's political campaign in support of its privatization agenda, it was a total failure. In a conjuncture marked by mass mobilization against the government's privatization agenda orchestrated by the National Federation of Indigenous Social Security (CONFEUNASSC) and joined by CONAIE and other popular sector organizations on the political left, the government's 'popular consultation' yielded a 60 per cent plus rejection of the government's proposal to privatize social security and to ban paralyzing strikes in the public sector. But this defeat did not deter the government from proceeding with other parts of its neoliberal agenda. Nor did it slow the rhythm of class struggle and violent confrontations over the land question. On the one hand, both the Chamber of Agriculture and CONAIE presented the government with alternative legislative projects related to this question. On the other, CONAIE stepped up its tactic of mass mobilization. To force the government to negotiate the demands of the indigenous movement and to engage it in a process of dialogue, indigenous communities and peasant organizations across the country mobilized a series of direct actions including protest marches and '*cortas de ruta*'—the blocking of highways and transportation roads, a tactic used with considerable success (political impact) by indigenous peasants in Ecuador and elsewhere throughout the 1990s and more recently by the movement of unemployed workers in Argentina. The press saw these direct actions as the radicalized form of another possible 'uprising.' Although this did not materialize, it was clear to everyone, including the official press, that 'the antagonistic position of the big and medium-sized landowners (on the one hand) and the *minifundista* peasants and landless workers (on the other) constitute an explosive conjuncture that can produce a very grave situation in the future if not resolved within a framework of dialogue and social justice' (*El Tiempo*, June 16, 1993).

In December 2003 the government introduced a state modernization law eliminating price controls and soon thereafter a law (March 2004) which opened up and liberalized the domestic market and then another law (April 2004) designed to 'modernize' the agricultural sector—a law that was regarded by CONAIE as a 'death project' for indigenous peoples, something similar to what NAFTA represented to the indigenous peoples of Chiapas.[10] The 1994 Agrarian Development Law was intended to put an end to the land reform program initiated in 1964 in the form of transfers of land to the tiller (peasants/indigenous communities).[11] However, in the immediate context of 1994 it resulted in a heightened level of class conflict over land.

1995 saw another turning point in the popular movement mobilized and led by CONAIE but orchestrated via the formation of the *Coordinadora de Movimientos Sociales* (CMS), a loose coalition of 34 labor and social organizations as well as indigenous groups and organizations that included FENOCIN, a leftist grouping of indigenous peasant communities and Blacks from the coast that led the indigenous struggle in the 1970s—and CONAIE. The advance in the popular movement, as well as a shift in emphasis from the land question to the issue of indigenous 'rights' and state reform,[12] and a sharper focus on broader national issues (opposition to neoliberalism, protection of the country's 'strategic resources,' etc.), occurred in a situation of deepening economic and political crisis. On the one hand, the country's economic problems, which most analysts attribute to the government's neoliberal policies which exacerbated rather than attenuated conditions of widespread poverty, assumed crisis proportions in 1995, a situation that reached its peak in 1999. Under these conditions of economic contraction and social decline, the political system entered another period of instability which in October saw Vice-President Alberto Dahik removed from office and that subsequently forced two presidents from office.

In this situation of impending economic and political crisis the indigenous movement, led by CONAIE, engaged its dual political project of a strategic alliance with the urban sector social organizations and the formation of *Movimiento de Unidad Pachakutik-Nuevo País* [MUPP-NP], a political movement[13] designed to allow the indigenous communities to participate directly in the system of electoral politics. In the 1980s the indigenous movement had participated in this system but for lack of a political instrument of their own they had to work through the party structure of the Ecuadorian left. In the context of the 1990 uprising, the government claimed that the uprising was orchestrated by the political opposition, forcing CONAIE to eschew any participation in the electoral system. In any case the leadership was of the opinion that the critical political dynamics of social change were to be found outside this system (Dávilas, 2004). Pachakutik, like the CMS, was conceived in 1995 but it was not until 1997 that it was given a solid organic structure with a national executive committee, provincial coordinating bodies, diverse commissions, etc.[14] This structure allowed CONAIE, through Pachakutik, to advance its own candidates for government office, without the need to support the candidacy of other politicians, even those who, like Freddy Ehlers in 1996, had very broad support of social organizations in the popular movement, advancing as he did an alternative popular program of economic and social policies.[15] The formation of Pachakutik also undercut the system of clientelism, a structural feature—and seemingly unavoidable part—of electoral politics.

Notwithstanding the consolidation of an indigenous political movement,[16] the major political event of 1997—namely the ousting, in February, of Abdal Bucaram's four-month old government—was brought about not by electoral means but by mass mobilization. In this regard, the ousting of Bucaram can be compared to the similar event in January 2000 when the Muhuad regime was overthrown temporarily (for four hours) replaced by a triumvirate that included Antonio Vargas, CONAIE's president at the time, as well as Lucio Gutiérrez, a leading representative of a group of young army officers that had moved against the government. The ousting, in December 2001, of two presidents in Argentina and the overthrow and forced exile (to the US) of Bolivia's De la Lozada, in October 2003, provided similar lessons to the popular movement: that mass mobilization provides popular sector organizations with a more effective means of regime change than the system of so-called 'democratic' elections, which, from the perspective of long experience, has proved to be full of traps and pitfalls, dominated as it is by the 'political class' (the 'establishment') and its machinations. We elaborate on this point below.

The populist politician Abdal Bucaram was a tragicomic figure in the convoluted history of Ecuadorian class politics. He represented the coastal (Guayaquil) commercial bourgeoisie in its longstanding factional dispute over national politics with the *sierra*'s landed oligarchy represented by the traditional parties that have tended to dominate both the legislative and executive branches of the government. The push to remove him from office was precipitated by his efforts to extend the package of neoliberal reforms and austerity measures introduced by Léon Febres Cordero in 1984–88, continued in the subsequent social democratic regime of Rodrigo Borja (1988–92), and further extended by Sixto Durán Ballén in his regime from 1992 to 1996 (Vicuña Izquierdo, 2000). Febres Cordero had turned Ecuador's national economy toward a neoliberal model of capitalist development with the adoption of an IMF-mandated stabilization program of austerity measures and a process of structural adjustment in the form of financial and trade liberalization. Borja, in response to the Washington Consensus as to the required package of economic reform measures, sought to insert Ecuador into the recently announced and heralded process of 'globalization' via policies of trade liberalization and labor market reform, to create thereby a more flexible regulatory regime vis-à-vis the labor process—*la 'flexibilización laboral.'* The contributions of Durán Ballén in regard to this reform process was both to deepen the IMF's program of austerity measures and extend the neoliberal 'modernization' reform program via a policy of privatization, turning over strategic areas of the economy (oil, electric generation) to the private sector. Because of the offensive launched by Durán Ballén against labor and the labor movement at the very beginning of his regime, the privatization of public sector enterprises became the central issue of class conflict and remained so until 1998. The public sector at the time (1992) included the most powerful unions and the country's biggest enterprises—Petroecuador, Inecel, and Emetel. According to CAAP, which monitors the level of social conflict, of 1000 significant social 'conflicts' from 1996 to 1998, 30 per cent originated in the public sector and revolved around the issue of privatization (*Equipo de Coyuntura del CAAP* 1998). This was more or less the same share of total social conflicts as Sánchez Parga (1993) found for the labor movement in the 1980s in the heyday of Ecuador's labor movement. Thus, it is not surprising that one of the major means proposed and used by the government (in its 1998 constitutional reform program) to secure '*gobernabilidad*' (governability, governance)[17] was to take away the right (and thus the capacity) of workers in the public sector to paralyze vital services to the public.

When Bucaram assumed the presidency in August 1996, Ecuador was well down the neoliberal path and in the throes of a protracted economic crisis, which had generated a mounting level of social discontent with conditions of poverty that afflicted 31 per cent of the population in 1988, 56 per cent in 1995 (76 per cent in the rural areas) and, after the adoption of dollarization (an extreme form of neoliberal development) in 2000, 68.8 per cent (Larrea, 2004: 50). Policies pursued by a succession of neoliberal regimes had drastically reduced the share of workers (wages) in national income from 31.9 per cent in 1980 to less than 13.6 per cent in 1990 and less since, one of the lowest in Latin America (Vicuña, 2000: 76). The purchasing power of the miserable (below-poverty) level of income or wages received by the majority of Ecuadorians experienced a 22 per cent fall from

1988 to 1992 and another 73.6 per cent from 1995 to 1999 (Vicuña, 2000: 126–7), a situation exacerbated by a general decline in the level (and coverage) of government expenditures on social programs.[18] In addition, government policies in the 1990s brought about a dramatic weakening of the labor movement vis-à-vis the capacity to negotiate collective contracts for higher wages and improved working conditions (SAPRIN Ecuador, 2004: 69–75). The same policies also generated a huge outmigration—up to 700,000, it has been estimated, since 1998 alone (Vicuña, 2000: 27). Today, remittances from these migrants constitute the second largest source of national revenues after the earnings generated by the export of oil.

In the specific conjuncture of economic and political conditions under which Bucaram assumed office, the national economy was already in a critical state and in no position to withstand the pressures of a controversial plan to fix the rate at which the *sucre* could be converted into dollars, a plan that would emerge a few years later in the dollarization of the economy (Larrea, 2004).[19] Bucaram took over the government in a situation of incipient and occasionally disruptive social conflict in reaction to the government's neoliberal policies. But in retrospect it would seem that the most critical factor in his government's downfall was not so much the parlous state of the economy and the conditions of a broad and deep class struggle as the internal political conflicts within the ruling class, elements of which possibly helped engineer the heightened level of political conflict that shook the Bucaram regime[20] and certainly took advantage of the popular mobilization against the government (Vicuña, 2000: 116).

One of the few neoliberal policy measures implemented by the government in 1996 that was not at all unpopular was the decentralization law, which responded to pressures from the World Bank and other international organizations. It was designed to address the government's fiscal crisis (by transferring certain responsibilities such as health and education to lower-level governments) while at the same time reducing the level of political conflict surrounding public policy at the national level. In theory, the policy of administrative decentralization, pioneered by the Pinochet regime in Chile, would not only eliminate a source of political conflict but establish the legal and institutional framework for a more equitable and participatory form of development, not to mention 'good governance' (World Bank, 1994; Blair, 1995; 1997; UNDP, 1996; OECD, 1997). But in practice, the policy induced and strengthened a process of local development within rural municipalities that were close to the indigenous communities in the highlands and in the Amazonian region.[21] This process, mediated by a host of NGOs,[22] was supported and financed by the World Bank, the IDB, and other international development organizations. However, although this process provided the indigenous rural population with an alternative to the direct action and mass mobilization strategy favored by CONAIE, it seems to have had relatively little impact on the political dynamics of the indigenous movement. A possible reason for this is that the leadership of CONAIE was—and still is—of the view that its political project could best be realized via a combination of diverse strategies and tactics, including mass mobilizations when necessary, negotiations, and dialogue with the government where possible, and both participation in the electoral system and the alternative local development projects financed by the World Bank and the Inter-American Bank. As of the 1990 uprising, according to Marcelo Córdoba, an indigenous leader, the indigenous movement has been based on two major demands, indigenous 'rights' (to land, territory, etc.) and 'development' (cited in Rhon Dávila, 2003).

The move against Bucaram was apparently triggered by a general strike orchestrated by CUT and a coordinated coalition of social organizations. However, it did little to change the policy orientation of the government, which continued on the same neoliberal path, with the implementation of the law of economic transformation ('TROLE') I–II, an omnibus program of diverse measures designed to achieve structural adjustment of the Ecuadorian economy to the requirements of the global economy—and, it can be said, to benefit small groups of economic and social power within the country as well as the guardians of foreign capital. In regard to the latter, the government from 1997 to 2000 paid over $25 billion in interests alone, triple the export earnings over the period and representing up to 45 per cent of central government expenditures and 10 per cent of the GNP (Larrea, 2004: 83; Vicuña, 2000: 1, 16).[23] In making these payments the government not only shortchanged its system of social and development programs but cut back any plans for productive investment, mortgaging any hopes for the country's future and for escaping economic crisis. In addition to these direct economic and social costs the series of agreements made with the IMF to bear these costs included a commitment to implement policies that were bound to make things worse—and did.[24]

The economic and social costs of this process of 'adjustment' were exceedingly high by any measure as the economy spiralled into its worst crisis to date and the 'poor,' some eight million according to the World Bank (half of this population 'indigent,' that is, with 'earnings' of less than $1 a day), were compensated with '*bonos de la pobreza*' (welfare payments) of $6 a month, recently adjusted to $10.50 (Vicuña, 2000: 176). Per capita production declined by some 30 per cent from 1998 to 2000, with a corresponding decline in per capita income (Vicuña, 2000: 174). Under conditions of massive decapitalization and the destruction of capital invested and deployed in diverse sectors of the economy, thousands of small and medium sized firms went bankrupt or disappeared.[25] Unemployment soared to record heights (from 8 per cent to 17 per cent), as did underemployment (up to 60 per cent), while the population living and working in conditions of poverty doubled from 1988 to 2000 (*Fundación José Peralta*, 2003). These conditions were exacerbated by an inflation rate that climbed to 91 per cent in 2000 and the virtual collapse of the banking system, provoking a turn toward a policy of dollarization—exchanging 25,000 *sucres* to the dollar, representing a devaluation of 360 per cent (Vicuña, 2000: 175; Larrea, 2004: 36). This policy, together with the conditions of TROLE (law of economic transformation) I and II, two omnibus packages of stabilization and adjustment measures, in turn sparked a broad insurrectionary social movement, an uprising that on January 21 would bring down the government.[26]

Dollarization had the immediate effect of reducing by at least 40 per cent the value of wages and the incomes received by the majority of the population. The economic conditions of this effect were as dramatic as its political effects. For the popular movement dollarization was the straw that broke the camel's back—the last straw as it turned out. Within days of the currency conversion law, CONAIE mobilized a march on the government, lighting the match of a nationwide conflagration—an uprising that found broad support in the popular sector, generating thereby an explosive combination of oppositional forces that came to a head on January 21, 2000, bringing down the government.

The diverse conditions that combined to produce the uprising and overthrow of the government, marked by *six days* of active resistance and heated struggle, have been subject to considerable analysis, mostly retrospective in form.[27] Piecing together diverse accounts of these and related developments (see, in particular, *Bulletin ICCI*, various monthly issues, February 1998 to December 2000) is not easy, but it seems that in addition to the objective conditions generated by the neoliberal policy measures associated with TROLE, there were at least three other factors. First, there was the situation of serious discontent within the armed forces, with a group of disgruntled officers concerned about salary and institutional integrity issues.[28] In this situation a group of middle-ranking officers, headed by 'Lucio' (Colonel Gutiérrez), entered into negotiations with Antonio Vargas, the president of CONAIE, and the MCS. Secondly, the ruling class was seriously divided, unable to constitute a dominant bloc and formulate a coherent government program, with an important faction of 'the [highland] oligarchy' engaged in political machinations against 'the [Guayaquil] bourgeoisie' which created a favorable condition for insurrectionary action (Delgado Jara, 2000). Another and indisputably critical factor in the uprising was the existence of a highly mobilized mass of social forces within the highland indigenous peasantry in alliance with a broad coalition of urban-based popular sector organizations against the government's neoliberal agenda.

LUCIO GUTIÉRREZ AND THE POLITICAL DYNAMICS OF A NEW REGIME

January 21 2000 was a critical conjuncture in the history of CONAIE and the popular movement in Ecuador. Another critical conjuncture, at least vis-à-vis the relationship of the social movements to the state, involved Gutiérrez's electoral campaign for the presidency and his eventual election. A critical factor in, and indispensable condition for, this political development was the support given to Gutiérrez by CONAIE as a social movement and Pachakutik as an electoral apparatus. Gutiérrez himself had no political base except for his hastily formed 'Sociedad Patriótica' (Patriotic Society Party—PSP). In this situation Gutiérrez opted for an alliance with Pachakutik and MTD, an urban-based leftist political grouping, exchanging for their electoral support an agreed number and distribution of positions within the government. The CMS, a broad leftist-oriented coalition of urban social organizations, was excluded from this alliance largely because Gutiérrez had no intention to saddle his government with a commitment to a popular policy program. It seems that notwithstanding a signed agreement on principles,[29] neither Pachakutik nor the MTD made their support conditional on such a commitment.

An effective political campaign and regime need a solid social base and are predicated on a coalition of social and political forces, but Gutiérrez constituted his electoral alliance on the basis of his virtually blank check agreement with Pachakutik, undoubtedly the decisive factor in his electoral success. In retrospect it is clear that Gutiérrez had a different conception of the social base and political support that would be needed to support his government. This base and support would be sought and found within the political class—the oligarchy to be precise.

With Gutiérrez's first-round victory, the Ecuadorian political left celebrated the anticipated rise to state power of the country's first military officer to do so via democratic elections rather than a military coup. In the event, wrote Gerard Coffey (2002, p. 4), 'the most generalized feeling is that of hope of change' Gutiérrez himself observed, 'The country's privileged sectors have profiteered enough; it is time for the poor to hope for better days, and it is to this end that I will dedicate my efforts' ('*Tintají conversa con el presidente Lucio Gutiérrez, Tintají*,' No. 16, 2003, p. 3). A group of Latin American and European leftist intellectuals published a letter hailing the electoral victory of Gutiérrez as 'part of a long process of struggle and resistance' ('*Carte al presidente lucio Gutiérrez*,' *Tintaji*, No. 16, 2003, p. 12).

At the end of the first electoral round, the left seemed to have shared to different degrees this optimism in regard to a Gutiérrez government. In this political imaginary Gutiérrez was part of a new emergent political axis, a shift in the political tide, together with Lula (Ignacio da la Silva) in Brazil, Hugo Chávez in Venezuela and Evo Morales in Bolivia (see, for example, the cover of *Tintají*, July 2002). This widely adopted position on the left originates in a false homology made between 'January 21, 2000'—considered as a single process—and the alliance of Gutiérrez's with Pachakutik. For example, it was noted by many observers and analysts that 'Lucio Gutiérrez is nothing without January 21'. Similarly an MTD leader argued that the candidacy of Gutiérrez as a political project 'was born with the popular uprising of January 21' and thus opposed by the dominant class in its entirety (*Tintají*, No. 28, p. 2). The idea was that the electoral victory of Gutiérrez signified the installation of the 'truncated government of 2000' (Lucas, 2003, p. 2). In this political imaginary Gutiérrez emerged victorious in an electoral contest with an oligarchy for state power, and he did so in the political representation and personification of the popular struggle, particularly as regards the indigenous question. And the leadership of Pachakutik shared this political imaginary.

This imaginary, however, did not take long to evaporate. Gutiérrez had already visited the US between the first and second round of the electoral process to assure Washington that he was 'its best friend in Latin America' and within four days of assuming the presidency his government signed an agreement with the IMF that violated his 'understanding' with Pachakutik (and CONAIE), namely that he would oppose the formation of the Free Trade Agreement for the Americas (ALCA in its Spanish acronym); reject the privatization of the public enterprises in strategic areas such as oil production and the generation of electrical power as well as pension funds and other vital social services; and turn away from the model (neoliberalism) used by a succession of governments since 1985 to shape national policy.

As for Pachakutik, despite all sorts of signals and the evidence of a growing number of policies made by Gutiérrez in a neoliberal mold and on the basis of overtures to and pressures from the US, it took the leadership up to six months to draw the inevitable conclusion—that it had been used, if not manipulated; that, in effect, Gutiérrez had 'betrayed' the popular movement. It would appear that the leadership drew this conclusion only reluctantly and only then because of a growing disenchantment and outright anger within the social base of the movement. Pressures arising from this disenchantment and opposition to the government's policies, as well as the complicity of their own leadership, compelled CONAIE to hold a national convention (Congreso Nacional III), in September 2003, to give an account of their political practice and to evaluate the participation of Pachakutik in the government. The CONAIE leadership at this convention was instructed to withdraw from the government forthwith and move into opposition. In short order all but a small number did indeed withdraw, creating a new conjuncture and opening up a new chapter in the history of CONAIE—a process of rebuilding the lost confidence of its members in the leadership and the reconstitution of this leadership.

In the context of this development the nature and meaning of the Gutiérrez regime became the object of an intense analysis and political evaluation. At the time of writing (March 2004) this process of analysis and evaluation is by no means over. Indeed, what we have is a variety of theoretical and political perspectives—if not confusion and disorientation. Nevertheless, it is possible to come to a number of conclusions, albeit tentative, in regard to subsequent political developments, such as the 2002 electoral campaign and the Gutiérrez regime.

The 2002 elections

The political forces behind the January 21 uprising did not endorse Gutiérrez's electoral campaign two years later *en bloc*. Between the complex dynamics of the uprising and Gutiérrez's bid for the presidency there is a huge gulf and substantive differences in political dynamics. Those who did not—or do not—see this tend to view 'January 21' in Gramscian terms as a constitutive element in the formation of a contra-hegemonic bloc of forces in resistance and leftist opposition—the constitution of a popular, or dual, power.

However, 'January 21' could just as easily be seen as a combination of diverse political processes that included a *coup d'état* in favor of Noboa Bejarano, the successor to Mahuad to state power. Those like Saltos (2001), who at the time wrote and spoke of a 'dual power' that was represented in the formation of a 'popular parliament', might be asked: can we conceive of 'popular power' in the context of subsequent developments? As Saltos saw it, January 21 did not represent a strategy aimed at the conquest of state power. Rather, it pointed toward the construction of a power 'from below' arising from the formation of a new 'historic bloc' formed by a convergence of the indigenous movement and a broader social movement against neoliberalism (Saltos, 2001). The result of this convergence was what Moreano, following Marx, termed the 'Quito commune.'

In this analysis, the failure of the uprising to achieve and consolidate state power resulted not from the betrayal of the military high command, machinations within the 'political class,' or behind-the-scene interventions of the US embassy—all indubitably present—but from the fact that the indigenous movement did not have a 'will to power'—that CONAIE did not want 'power' but saw itself as obliged to assume responsibility for it or alternatively, that CONAIE did indeed have a 'will to power' but—as argued by Latacunga (interview, September 17, 2001)—it was not prepared for state power. In this interpretation, shared by some participants and observers close to the indigenous movement, the significance of January 21 was more symbolic than political (an announcement of what is to come).

On the other hand, other analysts such as Pablo Dávilos (2003), who is also close to the indigenous movement (albeit as a '*mestizo* intellectual'), observe and argue that the indigenous movement is oriented toward state power—to a takeover of the state as the best and necessary political instrument for bringing about social change. In this interpretation January 21 implied a central shift in the strategy of the indigenous movement away from its self-construction as an 'antipower' ('*contrapoder*') toward state power. Other analysts agree with this interpretation but trace this strategic shift back to 1995—to the formation of Pachakutik. The issue in this view is how to achieve power—by electoral or constitutional means, per Pachakutik as an electoral apparatus and political movement, or by means of a strategy of mass mobilization, per CONAIE as a social movement.[30]

To deconstruct the meaning of Gutiérrez's presidential candidacy and electoral campaign it is necessary to look more closely at what the left, in its hurried and premature judgment, ignored: the political antecedents and orientation of both Gutiérrez and his running mate, Alfredo Palacio; and, more significantly, the fact that for the first time since the restoration of 'democracy' in 1978, the left did not put forward its own candidate.[31] First, in regard to Gutiérrez, his career spent entirely within the institution of the armed forces and no connections whatsoever with the political left, he stated in categorical terms some ten months before the elections that 'the *Sociedad Patriótica* is not a leftist movement, nor are we looking for one of its candidàtes' (Delgado 2003).[32] Delgado also reported on Gutiérrez's confessed sympathies for Pinochet and Chang Kai Sek. As for Palacio, a Boston-born 'entrepreneur' and a member of Ecuador's oligarchy, he as Minister of Health had participated in the rightwing government instituted by Sixto Durán Ballén. Palacio was also very close to Guayaquil's power structure, with solid credentials in regard to his predilection for neoliberalism and support for the policy of privatization instituted by the government of which he was a part. Apart from these considerations it is also useful, if not important, to remember that Antonio Vargas, president of CONAIE on 'January 21,' and a member of the short-lived (four hours) triumvirate and 'government of national salvation' was also a presidential candidate in 2002. In addition, for what it might be worth in regard to the formation of a broad anti-hegemonic bloc, the '*Partido Socialista—Frente Amplio*' (Socialist Party—United Front), which was part of the popular movement against the Mahuad regime, withdrew its support from Gutiérrez in 2002, electing to support instead a center-right candidate (Roldos). So much for an anti-hegemonic bloc of oppositional forces.

Gutiérrez's electoral campaign and successful bid for the presidency was based on the alliance of the '*Partido Sociedad Patriótica 21 de Enero*' (PSP) with Pachakutik (MUPP) and, to a lesser extent, with the Popular Democratic Movement (PPD), mobilized in the urban centers by the Communist Party. In addition, the Gutiérrez campaign had the support of a broad array of organizations on the left, including the CMS, the *Seguro Campesino* (Peasant Social Insurance), CEOLs, the CTE (Confederation of Ecuadorian workers), and FENOCIN, a federation of indigenous peasants and blacks organized by the Socialist Party. These organizations constituted a local political movement and electoral apparatus for Gutiérrez, although they did not reach the dimensions, or have the organic structure, of a counter-hegemonic bloc in a Gramscian sense. The alliance was secured with an agreement signed with Pachakutik on December 27, 2001 by Gutiérrez and officials of the so-called 'alternative local governments' which were formed by CONAIE *after* January 21. But the political base and center of the alliance was Gutiérrez's own party apparatus, the PSP, which to all accounts was little more than a means of bringing together into Ecuadorian politics the military officers involved in the January 21 uprising/coup. This group of officers constituted the original nucleus of PSP, but it was later joined by a broader group of retired officers from the armed forces and police services, mostly in the middle ranks of the institutional hierarchy, several wealthy ex-military business operators, and groups of petit-bourgeois 'professionals' (Ibarra, 2002: 28).

In sociological terms, these groups are largely drawn from diverse intermediate sector (middle-level) organizations and, strictly speaking, do not constitute a social class. At best they can be viewed, as Gramsci did, as members of various 'auxiliary' or 'instrumental' classes, or in Marxist terms, as part of the petit-bourgeoise in its diverse sectors— intellectual, business, professionals, institutional. In political terms, the provincial leaders of the PSP have been generally characterized as center-right and right in their ideological-political orientation and indeed their ties (as former activists) to center-right and rightwing parties have been noted by political analysts. However, it would appear that the PPP also includes former activists and members of several center-left and leftwing parties, although they are in a distinct minority. The critical factor vis-à-vis the role of these politicians in the conjuncture of the elections is the close connections and associations that many of them seem to have with the indigenous communities and peasant organizations in the countryside. The provincial leaders of the PPP played a critical role in the construction of the Gutiérrez–Pachakutik alliance but, he argues, at the center of the alliance was a group of individuals who had a structural connection to the leaders of the indigenous local communities at the level of shared social status, but at an ideological level treated this population as a subaltern group, viewing them in racist terms as part of an inferior civilization. In other words, the *relation* between Pachakutik as a political movement and this social base mobilized by Gutiérrez and his political cadre had a structural dimension and did not merely reflect the opportunism of a temporary political alliance within a specific historic conjuncture—*January 21*.

Gutiérrez's political discourse over the course of his electoral campaign was not anchored in any specific ideology or programmatic plan for governing the country. It had no ideological consistency or coherence and it shifted with the changing winds of political opportunism. In the campaign leading up to the first round of voting, Gutiérrez appealed to the political left with a discourse that emphasized opposition to payments on the external debt, the privatization of public enterprises, dollarization of the economy, the Free Trade Agreement for the Americas, US military bases in Manta, regionalization of Plan Colombia and the '*paquete*'—or, in stronger language with reference to its devastating impact, '*el paquetazo*'—the package of neoliberal reforms presented to the Congress for action by every government since 1985 and *the* source of the social conflict that has accompanied these governments. The only surprising feature of the political dynamics associated with this discourse was the apparent gullibility of the left who, for their own reasons or misplaced optimism or belief in their mobilizing capacity, bought into Gutiérrez's promise to 'form a government against neoliberalism'—to 'bring about a different country with less injustice and more equity' (*Boletin ICCI-RIMAY,* 2003: 4). Between the first and second round of voting, Gutiérrez visited Washington to convince the State Department and Wall Street that he 'was the US's best friend in Latin America' and that he would stay the course—stick with the policies that he denounced in his leftist discourse. At the same time, on his return to Ecuador, and in anticipation of the second conclusive round of the election process, he redoubled his efforts to secure electoral support, and campaign funds, from different social and political sectors of the electorate, the political class, and the media. In this campaign he managed to bring on board, among others, the Liberal Party, the Freedom Party, several powerful media and communication networks, and a host of organizations joining the bandwagon of campaign finance.[33] The deals and political pacts made in this process are, of course, not public, sealed as they are behind closed doors, only becoming apparent, and then only to a very limited degree, over the course of the subsequent government administration.

Notwithstanding the agreement in principle between Gutiérrez and Pachakutik, it would seem that the government's program of action was only formulated, and then merely in its general lines, between the second round of the electoral process and a few days before the formation of his government. To generate a social consensus on a program of action the PPP–Pachakutik alliance set up a process of 'dialogue and national unity'—a series of policy forums, from November 8 to January 25, which was designed, in theory, to promote 'a new participatory form of politics (policy formulation and decision making).' However, by the time that Gutiérrez took office and began to form his government, on January 20, this 'participatory process' was incomplete, yielding no concrete plan. That is, in the design of the specific direction that his government would adopt Gutiérrez would not be constrained by a predesigned program to which he was obliged to commit. Thus, he could promise all sorts of things without having to deliver on them, and probably without the slightest intention of doing so. And this was particularly the case in regard to CONAIE and Pachakutik, whose leadership had been effectively accommodated and muffled in terms of its capacity to advance either its central political project (the institution of a pluri-national state) and its broader policy program.

On the one hand, Gutiérrez had the support not only of Pachakutik and the MPD, his official electoral allies, but also a host of other leftist organizations constituting a local electoral apparatus in communities across the country. On the other hand, not one of these organizations, Pachakutik included, presented Gutiérrez with a concrete plan for managing the economic crisis in the short and immediate term.[34] Each of these political organizations did present Gutiérrez with a set of strategic issues and general principles but none had a plan. On November 8, 2002, beyond some value pronouncements and general expectations, the sole basis of the electoral alliance that led Gutiérrez to state power was an expectation for some positions in his government.

Gutiérrez's policies

Without an operational plan or any legislative measures to present to the National Congress, and with the governing coalition in a minority position within the legislature, Gutiérrez and Palacio were installed on January 20, 2003. The first sign of the direction that this government would take, and its essential form, was the construction of a team of senior ministers made up predominantly of well-known neoliberals with ties to the ruling class—to private firms that had promoted dollarization, a further 'opening' toward global capital and the need for further adjustments, including the privatization of the remaining state enterprises. The naming of the economist Mauricio Pozo, former vice-president of the *Banco de la Producción*, as Minister of Finance, and other monetarists and neoliberals to head government departments in strategically important areas, indicates that Gutiérrez's coalition partners, Pachakutik and MPD, had no say in or veto over the distribution of government positions—and thus the policy orientation of the government.

The leftist parties in the government coalition, MUPP-NP and MPD, were assigned five *ministerios* but only one—agriculture—entailed the capacity to implement alternative policies. Pachakutik (Luis Macas) was assigned the Agriculture portfolio while MPD (Edgar Isch) was given the new portfolio in the area of the Environment. However, both ministers would find their decision-making capacity severely constrained by the general policy and institutional framework of the model that guided government policy overall. Macas, although in a critically important department (Agriculture) and supported by a highly competent complement of vice-ministers and staff, found himself in a difficult position. He was able to proceed in a relatively short period of time with a process of titling—giving indigenous communities and peasants legal title to their land—strengthening thereby a class of smallholders within these communities, giving them access to sources of credit as well as enabling them to buy and sell—mostly sell—their property. But this policy was structured by legislation designed by the World Bank to 'modernize' agriculture, that is, introduce a 'market-assisted' form of agrarian reform, which, unlike the state-led agrarian reform programs of the 1960s and 1970s, is specifically designed to generate a productive capacity and increase productivity in the sector while shaking out the vast majority of marginal producers.

Another early sign of the nature of the government was Gutiérrez's signing of a letter of intent with the IMF only four days into his government.[35] This agreement violated key elements of the implicit understanding with his coalition partners as well as his campaign rhetoric and an agreement signed by Gutiérrez and representatives of a host of popular and labor organizations. The legislative measures and economic policies subsequently implemented by the government on the basis of this agreement have been characterized as 'the most orthodox expression of the current of thinking that has dominated Latin America over the last two decades'. In line with this thinking the Gutiérrez regime prioritized its 'obligations' to the IMF and other international financial institutions (debt service, interest payments) over a policy of productive investment to reactivate the economy. Thus, for example, the Minister of Agriculture, Luis Macas, was not part of the government's Productive Reactivation Commission. Also, in the 2003 budget submitted to the Congress, the government assigned only 2 per cent to agriculture but 36 per cent to debt service. In addition, the government has legislated that any 'surplus' generated by the export sales of oil over and above a price limit of $26 per barrel (the current world market price is $38) has to be set aside for debt service. Another such policy was for the government to buy back a part of its external debt, raising the value of these debt instruments in the process (leading the government to pay 60 cents on the dollar when the secondary market for these debt instruments had been 25 cents or less).

In the same context, in clear violation of the understanding reached with the social movements, the government in its short term in office advanced legislation in the direction of privatization in the strategic areas of oil production, electricity generation and telecommunications, by assigning to foreign capitalist enterprises in these areas critically important and valuable 'concessions'—to drill for oil, generate power, and set up networks, etc.—contract revisions, and proposals for joint ventures. Not only did the government respond to the demands of the 'private sector' (foreign and domestic capital) in regard to this policy, but it launched its policy in these strategically important areas with a systematic attack on public sector workers, blaming them (their high wages, self-assigned 'privileges,' corruption, abuses of excessive union power, etc.) for shortfalls and problems in production, as well as imposing a two-year wage and salary freeze.[36] In this regard in his aggressive campaign against public sector workers, and maligning organized labor, he has adopted and repeated the anti-labor discourse of the conservative Febres Cordero. In addition, Gutiérrez's 'accounting' of his first eight months in office (Ecuador, Presidencia de la Republica, 2003), is replete with references to the corruption of public sector labor union leaders. And, as part of his campaign to create the political conditions for a policy of further privatization in nationally strategic or politically sensitive areas such as oil production and social security (pension funds), he has supplemented his campaign against public sector workers with a policy of intimidation and repression, firing oil workers in the union leadership and bringing about a legal order for their arrest and detention.

The 'accounting' given of his government over the first eight months—now a full year—is in dramatic variance with the accounting made of it by the organizations in the popular movement, and on the left within and beyond this movement. The two major organizations to provide Gutiérrez with the social base for his electoral government—MUPP and MPD—have both abandoned the government in disgust with what some prefer to interpret as Rodriguez's 'move to the right' or his 'opportunism' *('oportunismo de un advenedizo')* but that many others, particularly in the indigenous movement, view as a fundamental 'betrayal' of their trust (*Tintají*, No. 28, 2003: 3). In the advent of this perceived betrayal, and on the basis of a collective decision made at the First Summit of the Nationalities, Peoples and Alternative Authorities (*I Cumbre de las Nacionalidades, Pueblos y Autoridades Alternativas*) organised by CONAIE in June 2003, Pachakutik officially withdrew from the government, resigning from all government posts—five ministries, eight sub-secretaries, and the hundreds of positions in the sectional and local governments across the country—assigned to it. Only eight or so of the 450 Pachakutik members of the government in its various levels and divisions (central, sectional, and local) remained in place, reluctant to give up their 'positions,' but none of these is in a senior position.

Pachakutik (and thus CONAIE) rejoined the opposition but in a very divided state, with their mobilizing capacity seriously weakened. In this situation the response and strategy of CONAIE is to rebuild the movement from within and below, at the movement's base in the local communities where there is a groundswell of opinion and feelings of betrayal and disillusionment but also mistrust of their own leadership which had so brought them into such a misbegotten electoral and government alliance with Gutiérrez. On only one point is there a consensus: that Gutiérrez did not as much 'betray' the movement as 'use' it and that the leadership of the movement should be held accountable for setting back the indigenous movement—a social and political movement for 'a new country,' a 'pluri-national, democratic, participatory and intercultural state' in which the indigenous peoples and nationalities are accorded full access to the productive resources to which they are entitled, as well as social justice and human dignity, equity, respect for indigenous culture and traditions (CONAIE, 2003). In addition to a total withdrawal from the government, the First Summit also mandated the CONAIE leadership to present Gutiérrez with an alternative program of economic and social policies which included: (1) the immediate rejection of the latest letter of intent signed with the IMF; and (2) the abandonment or reversal of all 'neoliberal' (structural adjustment and stabilization) measures, such as privatization (in any of its modalities, particularly in regards to strategic resources and social security), payments on the external debt (to a limit of 15 per cent of the national budget), elimination of price controls and subsidies (in relation to fuel, gas, etc.), and labor market reform (definitively shelve its plan for labor flexibilization). The government plan approved at the First Summit (CONAIE, 2003) also included general and specific proposals to statify the national economy and restore the integrity of the public sector, and reorient the country's political economy, placing the economy in a direction of 'productive reactivation with equity.'

CHANNELS OF SOCIAL CONTROL

Relations of the state with the major social movements over the years have been based on diverse mechanisms that constitute what could be seen as a 'system of social control,' used to establish conditions that the World Bank has defined as 'good governance' (World Bank, 1989). In the case of the Ecuadorian state, the mechanisms of social control vis-à-vis the social movements have included:

1. Policies designed to integrate the indigenous population into Ecuadorian society and its political-legal and economic structures. The major institutional avenue for this process of integration over the years has been education—incorporating the indigenous population into the country's formal educational system—but in strictly legal-administrative terms this strategy of integration and assimilation, half-hearted as it was, can be traced back to the 1937 *Ley de Comunas*, which originated the paternalistic attitude which still fundamentally characterizes relations between the government and the indigenous population today. Because this strategy was—and is—predicated on a fundamental disrespect for indigenous culture and society, and the conditions of 'social integration' into what remains a racist society have been so profoundly unequal and inequitable, it has not worked well. Most indigenous communities have been content to remain immersed in their own world. Certainly, some degree and diverse

forms of integration can be identified vis-à-vis the situation that prevailed in the 1930s, but these can be attributed not to government policy as much as to the push–pull pressures and crisis condition of the indigenous economy, which has led to a large-scale and persistent trend toward outmigration and the abandonment of the rural world of the indigenous community.[37] Under these *structural* conditions indigenous society has experienced a debilitating process of 'de-communalization' (emptying of the traditional commune as a locus and habitat for sustainable livelihoods and as a space of social reproduction) and articulation of the indigenous communities with capitalist society (Rhon Dávila, 2003: 133). This process, Rhon Dávila adds, 'together with the disarticulation of the indigenous family and migratory escape, is an expression and local effect of global processes of [social] exclusion and disarticulation.' One of these processes derives from, and relates to, the neoliberal project to 'modernize' the system of agricultural production, i.e. subject it to the logic of capital accumulation, which in Ecuador, as elsewhere, is highly exclusionary as regards the indigenous communities and their increasingly marginalized production systems (Dávilas, 2004: 179).

2. Another mechanism of social control used by the government has been a tripartite corporatist social pact with private sector proprietors and the organized workers, setting thereby the form (dialogue and negotiation) and limits to the methods that can be used to settle issues of potential conflict in the negotiation of collective contracts. This tripartite pact, introduced by the Durán Ballén regime, was modeled on Mexico's experience, particularly as regards the *Pacto de Solidaridad Económica* (December 1987) and the *Pacto para la Estabilidad, la Competividad y el Empleo* (October 20, 1992). Under conditions of structural change and the political conditions created by these two pacts and associated labor reforms such as the requirement of collective bargaining by enterprise rather than by sector and increased flexibility in the contracting, use, dismissal, and downsizing of the labor force, the organizational strength and negotiating capacity of the labor unions have been significantly reduced.

These conditions are reflected in statistics that show a dramatic decline in union members and a general decline in the share of labor in national income over the 1990s. In the metallurgical sector, for example, the level of unionization fell by two-thirds in just three years, while overall today only 5 per cent of the labor force is unionized, compared to 40 per cent in the mid-1980s at the outset of

Ecuador's experiment with neoliberal reforms. What is surprising—or would be were it not for the working of the government's resolution of the relation of conflict between labor and capital—is the relative quiescence of the labor movement in the face of a major assault on its organizational and mobilizing capacity. The 'labor movement,' insofar as it still exists, is a shadow of its former self, seriously weakened and fragmented, with its leadership either accommodated to the system or unable to mobilize union members. To all intents and purposes (political mobilization against government policies supportive of capital and destructive to workers) in Ecuador today there is no labor movement.

3. In the public sector of what is left of the labor movement—its strongest part—the use of indemnization as a means of offsetting resistance to the expulsion of workers and the reduction of the labor force. This mechanism was first used by the Durán Ballén regime in 1992 and was notably used by Gutiérrez in his battle with the oil workers, in his move against and expulsion of close to 40 union leaders, who, the government argued, were grossly overpaid (up to $800 a month versus $150 for the average worker or $250–$300 for other 'qualified' or technically skilled workers). A reduction of public sector employees and workers has been a consistent element of the agreements that governments of the day have signed with the IMF, most recently just four days into Gutiérrez's election as president (the agreement was to cut 1,500 workers from the public sector which included 277,000 workers for the central government; 36,957 provincial and municipal ('seccional') government workers; 24,051 in decentralized enterprises; 16,155 employed by IESS; another 10,576 employed in diverse state enterprises such as Petroecuador or Entel; and another 69,955 workers who fall under the public service labor code and are thus considered 'public servants,' a total of 442,144 (AME, elaborated by PNUD; Navarro, n.d.: 113).

The policy of downsizing the public sector, together with the privatization of public sector enterprises, has resulted in a substantial decline of the public sector in numerical terms and also in its political weight. Since 1998 the capacity of public sector unions to mobilize workers and the broader public against the government's neoliberal agenda has been substantially weakened. The labor movement in this situation has had to cede leadership of the popular movement to CONAIE, the dominant organization within the indigenous movement.[38] The oil and electrical workers within what remains of the labor movement have continued their struggle, but have had to do so within the CMS and with decreasing political clout.

4. The use of various sorts of incentives to encourage class-based organizations to turn toward the use of the 'electoral mechanism' in their politics and political parties as an instrument in their struggle for state power. Several social democratic foundations, particularly Friedrich Ebert Stiftung, have been instrumental in turning class-based organizations, including social movements, toward electoral politics and supporting a reformist approach toward politics. On these dynamics see, among others, Milton Benitez (1992) as well as organizations such as ILDIS and FLACSO that have been heavily financed over the years by German Social- and Christian Democratic organizations and used to create a reformist culture and orientation within the intelligentsia. NGOs contracted and funded by USAID and the OECD have been similarly engaged in the 'democratic promotion' process—in selling the virtues of representative and participatory democracy ('democratic development')—and, in the process, serving as a form of 'colonization.'

In the same context both the government and organizations such as the World Bank and the IDB involved in the project of international cooperation for development have worked to promote and support organizations that do not have a class line but that in postmodernist terms valorize 'relations of difference,' cultural diversity, political tolerance, and pluralism—new social movements focused on issues of gender equality, environmental protection, and ethnic identity. The institution of the *Consejo de Desarrollo de los Pueblos y Nacionalidades* (Development Council of the Peoples and Nationalities) and the *Dirección Nacional de Salud Indígena* (National Direction of Indigenous Health) can be viewed in these terms. Even the institution of the *Universidad Intercultural de las Nacionalidades y Pueblos Indígenas* (Intercultural University of the Indigenous Peoples and Nationalities), a major achievement of the indigenous movement and a clear symbol of the government's belated recognition of the multicultural nature of Ecuadorian society, can be viewed as a means by which the government was able to defuse relations of conflict with the indigenous movement.

Until the mid-1990s, Ecuador's indigenous communities and peasant producers organized their politics, presented their demands, and mobilized in the form of a social movement, keeping their distance from the formally democratic political system of the state.

In 1995, however, there emerged a movement within CONAIE to form a political party—to integrate the movement thereby into the party system of Ecuador's 'political class' and advance the project of a pluri-national state from within. The indigenous movement found in Pachakutik a means of participating in the system of electoral politics, to present candidates for government office at the local, provincial, and national levels. At this level, and in this form, Pachakutik 'participated' in the elections of 1996, 1999, 2001, and 2002. And it thereby achieved political representation and presence in diverse political forums, particularly at the level of municipal politics but also at the provincial and national level. The decision in the mid-1990s to form a political movement in the form of an electoral apparatus shows how far the indigenous movement had changed. Hitherto the indigenous movement had always rejected the institutionality of the political system (liberal democracy). A number of arguments in this connection had been advanced over the years, including that there was no level playing field in which diverse groups could freely contest state power; on the contrary, the system was—and is—clearly controlled by 'the establishment' or 'oligarchy,' an economically dominant ruling class, much of which strictly speaking no longer takes the form of a landed 'oligarchy' but of a bourgeoisie.

The CONAIE leadership also had the idea that the most critical social conflicts between the indigenous movement and the broader Ecuadorian society could best be processed from outside the political system through non-electoral means (Dávilas, 2004: 34). Given this fundamental belief CONAIE retained its strategy to use a variety of tactics in its struggle and it retained its character as a social movement, organized to mobilize its social base in the form of direct mass action—marches, occupations, etc. In this non-electoral form CONAIE launched the *Intiraymi* uprising of 1990; participated in a mass mobilization against the privatization of the social security system in 1993—120,000 indigenous peasants were mobilized against the government in just one province (Azuay)—and mobilized a movement against the government's plan to 'modernize' agriculture (its 'agrarian law') in 1994; in 1995 it successfully mobilized a movement against a plebiscite organized by the government so as to allow it to proceed with its plan to privatize social security (60 per cent of the electorate voted against the government, a similar percentage to those who opposed the government's plan to outlaw paralyzing strikes in the public sector); in February 1997 it brought down the Bucaram government and mobilized massive support for a constituent assembly; in March and July 1999 it mobilized successful nationwide protests against the government's austerity measures and in September organized another mass mobilization— this time unsuccessfully. In addition to the mass mobilization (the 'Rainbow Rebellion') which brought down another government just a few months later in 2000, the indigenous movement in 2001 (January 27–February 8) mobilized against the Noboa government's attempt to reintroduce a new '*paquetazo.*' In each case the indigenous movement's success in stalling or slowing down the government's neoliberal agenda, if not forcing it to change direction, reflected its notable capacity for mass mobilization rather than its participation in electoral politics. A number of analysts of the indigenous movement, including CONAIE itself, look at the 1990s as a decade not only of class struggle but of significant 'gains' for the movement—gains attributed to its capacity for, and strategy of, mass mobilization.

Political developments subsequent to January 21 manifest a fundamental problem for social movements in regard to electoral politics, a problem experienced by movement after movement in country after country, and a problem internal to CONAIE as a social movement. The inevitable result of an electoral strategy appears to be the accommodation of the leadership to political dynamics that inevitably lead to the demobilization of the social forces of resistance and opposition accumulated by the social movements. Like most popular sector social and political organizations CONAIE learnt this lesson the hard way, having entered into an electoral alliance with Gutiérrez in his bid for the presidency in the expectation of political influence—and a number of positions within the government. As it turned out, Gutiérrez won the elections and Pachakutik, as well as a sector of the left represented by MPD, entered the government.

It took the political leadership of CONAIE some time to realize that it had made a strategic political mistake—selling out its membership by electing to join the government, attempting to advance its agenda from within, and thus suspending direct actions and mobilizations. Not all sections of the movement and its leadership share this understanding. In fact, the movement is divided between those like Cholango, ECUARUNARI's leader, who takes a class approach toward politics, and those like Vargas who are susceptible to the lure of opportunism baited by the government. And then there are the politicians involved with Pachakutik who are loath to draw the lesson that electoral politics is a dead end. Nevertheless, the lessons of the Gutiérrez government after six months in office allowed for no alternative interpretation. And in July 2003, after a national assembly convoked to the purpose, CONAIE officially broke with the government, enjoining its members to leave the government, resigning all government posts. This decision was not without its political difficulties but the vast majority did indeed leave the government, allowing CONAIE to reconstitute itself as a social movement—albeit in a situation of internal division and political weakness. In March 2004, at the time of this writing, CONAIE still finds itself in this situation, with an enormous distance between a divided leadership and a disarticulated social base characterized by widespread discontent, disenchantment, and disillusionment.

As a means of closing this distance (and recovering its erstwhile mobilizational capacity) in the wake of its official break with the Gutiérrez government, signaled by the withdrawal of Pachakutik deputies from their government posts, the CONAIE leadership has moved in two directions—a strategy of rebuilding the organization from the bottom up, at the level of the base community; and a new round of mobilizations aimed at removing Gutiérrez from power.[39]

5. The international development project was born, as it were, in the immediate aftermath of the Second World War as a foreign policy measure of the US government—the dominant power in an alliance of capitalist democracies—to ensure that the 'economically backward' countries that were emerging from the yoke of European colonialism would not fall prey to the lure of communism—that they would choose and stay the capitalist course for their national development. In the 1960s, however, the Cuban Revolution brought about a twist in this project. To prevent another Cuba, and undermine and thwart widespread pressures for revolutionary change, the US government and other governments of countries in the OECD initiated a strategy of social reform and integrated rural development, providing thereby the rural poor with an alternative to revolution, and enlisting the service of private voluntary associations (PVOs) to this end. Converted into nongovernmental development agencies, and contracted by the US and European governments as their private agents, these PVOs joined the army of Catholic NGOs ('Catholic Action') already in the field as lay missionaries, promoters of the virtues of reform and democracy, to provide what 'assistance' they could in the process.

In the 1980s the context of this project changed but the NGOs were called upon and enlisted to play a similar role as promoters of the twin virtues of democracy and capitalism—the marriage of free elections and the free market. The agenda set for the NGOs once again was to provide the rural poor with a reformist option to revolutionary or radical change; to turn them away from direct action and a confrontation of the structure of economic and political power toward micro-development projects within the local spaces available within this structure—development without change, democracy without social movements. In Ecuador, as elsewhere in the region, the World Bank and the Inter-American Development Bank assumed responsibility for creating the political conditions for 'sustainable human development,' namely participatory development' and 'democratic governance'—political order with minimum government, based on social rather than political control, a social consensus within civil society through a non-confrontational politics of dialogue and participation rather than collective action in the direction of revolutionary change. The essence of this approach toward social change (local development, no-power) is a reliance on the accumulation of social capital, an asset that the poor are deemed to have in abundance (Woolcock and Narayan, 2000; Bebbington, 2001). Rather than confronting the structure of economic power, indigenous organizations are encouraged to seek improvements in their lives within the spaces available to them within this structure.

In Ecuador, as in Bolivia, the indigenous (and indigenist) movement orchestrated by, and associated with, CONAIE has been the primary object of this strategy. The World Bank, the IDB, and other international development agencies such as USAID have poured millions of dollars into local development efforts, into support for alternative development projects that are in theory (but not in practice) initiated 'from below' and 'from within' civil society (Bretón, 2003). In this connection the World Bank set up and funded the Development Project of Ecuador's Indigenous Peoples and Blacks (PRODEPINE), co-opting thereby community leaders[40] and the then-president of CONAIE, Antonio Vargas, and effectively turning the indigenous movement away from its more radical demands as well as dividing it.[41]

In April 2004, at the time of this writing, the movement remains divided, its organizational and mobilizational capacity seriously weakened by a combination of factors—the accommodation of some leaders toward local 'alternative' development (micro-projects) and the prioritizing of its cultural/indigenist' agenda (recovery of its ancestral identity) over the issue of territorial autonomy and the 'land question'; an involution of its demand for a pluri-national state into the government's acceptance of the notion of Ecuador as a multi-ethnic 'society,' and, most importantly, entering into an electoral alliance with Gutiérrez. Today, there is virtual consensus in diverse political sectors that CONAIE is reaping what it sowed.

6. Under the neoliberal model instituted in 1982 and imposed in the 1990s (Carrasco and Marx, 1998: Larrea, 2004; SAPRIN, 2004), a policy pioneered by the Pinochet regime in Chile in the 1970s, took the form of the Law of Decentralization adopted by the government in 1996. The aim of this law and associated reforms was to create conditions for a participatory and sustainable form of local development—a matter of organizational efficiency, equity, and good governance (UNDP, 1996; Blair, 1997; Kaufmann, Kraay and Zoido-Lobatón, 1999). The paradigmatic form of such development can be found in Bolivia, but Ecuador's Decentralization Law of 1996 created a legal and institutional framework for a more decentralized and participatory form of development focused on the municipality or local government as a development agent ('the productive municipality') as well as a locus of decision-making (not in terms of macroeconomic policy but the implementation of the government's new social policy of 'solidarity with the poor').

On this institutional and legal basis a considerable number of CONAIE's community-based social organizations have been penetrated (NGOs as agents of a 'new colonialism')[42] and adjusted to the institutionality of the state—with increasing reliance on outside development finance and assistance. On the same basis the government has managed to institute a form of 'democratic governance' (participatory democracy and development). The secret of this form of organization is the replacement of centralized political control with a system of 'social control' based on the construction of a consensus within 'civil society.' In the case of Ecuador this system has worked well in certain contexts (see Cameron, 2003), but not so well in others, generating widespread conditions of political conflict as well as mass mobilizations in defense of the institutions such as the *Seguro Social de Campesinos* (Peasant Social Security). One of the major points of agreement reached by CONAIE with the Noboa post-uprising government (on February 7, 2001) was for the government to 'bring about a profound decentralization of the state via the Association of municipalities and the government's Special Commission on Decentralization'; and, in the same context, to support the coordinating body of 'alternative local governments' in their pursuit of 'alternative projects.'[43] The mandate given to the CONAIE leadership by the First Summit of Indigenous Nationalities, Peoples and Alternative Authorities in 2003 included policies designed to capacitate local governments to function as agents of economic and social development. Included here is a guarantee for an automatic transfer of sufficient financial resources from the budget of the central government to local governments.

7. An important means of social control instituted by the government some decades ago was to set up public institutions designed to subject indigenous communities and ethnic organizations to government policies—to incorporate them into the state. Examples here are the *Secretaria de Asuntos Indígenas* (Secretariat of Indigenous Affairs) and the *Instituto de Desarrollo Agrario* (INDA). Recent measures of this type include the institution of COMPLADEIN in 1997–98 and the *Consejo de Desarrollo de los Pueblos y Nacionalidades* (Development Council of Peoples and Nationalities). Another tactic used by the government over the years is to set up or support parallel organizations that the government can control to class-based organizations. This tactic has been particularly effective vis-à-vis social organizations and movements of indigenous communities and peasant producers. A case in point is the systematic and extended efforts of the government to set up and finance parallel organizations (AIEPRA, COIRA) in alliance with some evangelical groups in FEINE (*Federación de Indígenas Evangélicos del Ecuador*). The aim here was not only to provide an alternative to an organization (CONAIE) that the government could not control, but to offset its strategy of mobilization. However, as noted above, this strategy has enabled CONAIE to bring down two governments and continues to be the major counterweight to the government's continued efforts to impose its neoliberal agenda.

8. An effective method of social control and disarticulation of the class-based organizations and movements contesting state power is to co-opt or buy out the leadership, accommodating it to the government's social control agenda and efforts to divide the movement, disarticulating its social organization and demobilizing its forces of resistance and opposition. Examples of this tactic are legion but few have been so obvious and effective as in the case of Antonio Vargas, CONAIE's president at the time. It was evident that Vargas had his own agenda vis-à-vis the military establishment that had totally funded the event. The precise agenda was not clear,

but a contingent of the armed forces, headed by Gutiérrez, rallied to support the uprising and joined the triumvirate that briefly held state power for a few hours until the directorate, composed of Vargas, Gutiérrez, and Saltos, representing the CMS, was, in effect, betrayed by Gutiérrez's superiors in the military hierarchy, acting on instructions from the US Embassy. Vargas was subsequently defeated in his bid for another term as CONAIE's president but not before accepting an offer to head PRODEPINE, funded by the World Bank to the tune of at least $25 million. This was not a high price for the Bank to pay, considering the stakes and the desired outcome of turning the indigenous movement toward the local development option as well as, in the process, dividing the movement. In fact, numerous people within CONAIE and some of its closest observers are convinced that in exchange for favors and personal benefits, Vargas consciously accepted an assignment from the Gutiérrez government, with the support of the World Bank and the IDB.

9. Governments have at their disposal a variety of mechanisms for securing the compliance or social control of social movements. But when all of these options fail, governments of the day have recourse to instruments of coercion at their disposal. And, as a measure of last resort, they have generally shown themselves prepared to mobilize the repressive apparatus that is normally under the political control of the Minister of the Interior. And in its short history it is clear that the Gutiérrez government is no exception or slouch in this regard. Although there is no direct evidence implicating Gutiérrez himself, his year-old government has been responsible for numerous acts of intimidation and threats, and assassination attempts against the opposition, including, in December 2003, the president of CONAIE himself. Overall, the policy of the government—and in this regard the Gutiérrez regime is no different from previous governments—has been to decapitate, weaken, and demobilize social movements when and where possible, but to repress them if necessary.

CONCLUSION

After little more than a year in power Gutiérrez has left the country in worse shape than before. On this there is virtual consensus among all social and political sectors. In large measure this can be attributed directly to the government's abandoning of its pre-electoral commitment to reverse the neoliberal policies of previous regimes.

By courting, or perhaps succumbing to, the forces of global capital and the White House, and turning his government away from the popular movement, Gutiérrez effectively deprived himself of any possible means to reactivate the economy and foment a process of social and economic development. A critical factor in this is the prioritizing of the government's 'obligations' to foreign creditors vis-à-vis the external debt over and above its obligations to pay down the huge social debt accumulated under two decades of neoliberal policies. Other critical factors include (1) the fact that the social base of the government's policies is found in the economically dominant and political ruling class—the 'oligarchy'; (2) outside pressures on the government to integrate the economy into the globalization economy under conditions of structural adjustment; (3) the lack of a cohesive social movement able to articulate, politically represent the diverse sectors of Ecuador's civil society, and unify the forces of opposition on the basis of a Transition Program and a Plan for Action; and (4) the relationship of the state to the popular movement, particularly the indigenous communities that constitute the social base of the most powerful movement in the country, the most dynamic sector of the popular movement.

In regard to the relation of this movement to the Ecuadorian state the critical issues remain those of *social change and political power*—how to bring about the first and achieve the second. On these issues we conclude as follows. The political dynamics of the Gutiérrez regime can best be understood in terms of four political options: (1) a reliance by constituted social movements on their capacity for mass mobilization—to mobilize the forces of opposition to the state and its neoliberal policies; (2) the formation of a political movement with an electoral strategy—to contest state power *within* the formal liberal-democratic political system; (3) the constitution of an anti-power—an alternative counter-hegemonic bloc; and (4) a no-power strategy—to pursue the path of social (as opposed to political) action and local development, the implementation of externally funded micro-projects.

In an earlier political context (the 1960s and 1970s) this option was captured in the formula: *reform or revolution*. In the new context associated with the neoliberal model of capitalist development in the 1980s and 1990s, the option available to organizations in the popular sector of 'civil society' could well be expressed in the formula: *local development or social movements*. However, this option relates to the preferred or most effective form of social organization. In strictly political terms, the strategic option for social movements in the popular sector is for them to take power in the form of mass mobilization, or alternatively, to work within the formal institutionality of electoral politics and the associated political system. In other words: *revolution or reform.*

The political dynamics of the struggle waged by the indigenous movement over the last year of the Gutiérrez government reflects the long-established pitfalls associated with the system of electoral politics and representative ('liberal' or 'bourgeois') democracy. First, the system is dominated by what in Latin America is generally termed the 'political class' under rules and procedures that clearly favor the dominant economic class that it represents. It is this *reality* that led *subcomandante* Marcos in a different context (Chiapas) to argue for a new way of 'doing politics'—not to participate in any corporatist or paternalistic strategy, the system of client–patron relations or democratic elections, or to accept any positions within the government, one of several mechanisms used by the Gutiérrez government to compromise and co-opt the leadership of the indigenous movement, defusing the forces of resistance and opposition, and effectively dividing it from the social base of the movement. Participation in this system not only places popular sector organizations in a disadvantageous position vis-à-vis the capacity to mobilize the social forces at their disposal, but it subjects social movements to diverse mechanisms of social control available to the state.

The indigenous movement confronted the dilemma of political choice, *reform or revolution*, or, in organization terms, as *a sociopolitical movement oriented toward direct action/mobilization* or *participation in the political system*—to seek change from within, below or the outside—in a situation of an economic and political crisis. A crisis, as both Marx and Gramsci pointed out in different context, means that the old world is dead or dying but that the new world is stillborn or experiencing difficulties in being born. This is certainly the case for the indigenous movement. On the one hand, a process of neoliberal reforms, particularly financial liberalization, precipitated a momentous banking crisis that pushed the economy into its worst crisis ever. This outcome was experienced in different ways, with diverse permutations, in virtually every other country in the region, leading to a virtual consensus that the neoliberal model is not only non- or dysfunctional, to use the terminology of Carlos Slim, but, to all intents and purposes, is dead—and certainly dying. In Ecuador, however, this reality did not preclude a succession of governments from seeking to advance the neoliberal agenda—to stay the course and stick to the medicine prescribed by the IMF. What it did do is generate a broad popular movement against the neoliberal model of government policy—to help dig its grave. A clear manifestation of this was the transformation of the indigenous movement from its concerns with sectoral issues (rights, development, etc.) into an organization concerned with broad national issues: 'Nothing just for the Indians' ('*Nada solo para los indios*')—the motto of the 2000 uprising.

This strengthening and broadening of a popular movement against neoliberalism was a defining characteristic of political developments as of the late 1990s. On the other hand, the 'new' could not be born and has failed to come to light. The problem here is complex. First, as Vargas himself pointed out, in the quasi-revolutionary situation that existed at the end of 2000 the indigenous movement was unable to capitalize on its huge reservoir of popular power. Its mobilizational capacity (to take power) was not matched by an institutional or organizational capacity to exercise power; the movement lacked both an organizational capacity and an effective strategy. The power of the movement consists in its social and organizational base in the community—in the indigenous communities. However, neither CONAIE, the main force in the counter-hegemonic bloc in formation, nor its allies in the popular movement (CMS), were in a position to seize and consolidate state power. Various factors played into this, including a lack of organization (articulation between different forms and levels of indigenous organization) and a strategy for dealing with the existing power bloc and its control over the state apparatus. Nevertheless, CONAIE almost took state power, participating as it did in the construction of a short-lived (four hours) 'government of national salvation.'

The political dynamics of this brief interim period—which saw the transformation of a mobilizational failure into an uprising/takeover of the state apparatus—have been analyzed from diverse perspectives. But it is subsequent political developments, from the Mahuad regime to the Gutiérrez regime, that are critical to an understanding of the political dynamics of the indigenous movement—and that need to be reassessed, particularly as regards the relationship of CONAIE and Pachakutik to the incumbent government.

Viewing these issues retrospectively has led us to a straightforward conclusion that is entirely consistent with the evaluation made by many activists within the movement: that it was a serious political mistake to seek state power from within the system—to turn toward electoral constitutional politics and join the government. This much is obvious. Assessments of the state–movement dynamic in other contexts have reached the same conclusion. The problem is that this does not get us far. If neoliberalism is dead, why are its progenitors still standing? If it is dying, how best to dig its grave? With what means and where? And who will its gravediggers be?

As to the new world yet to be born, what are the dynamics of struggle involved? How are these dynamics to be advanced? Through what means and agency? To mobilize the forces of opposition and resistance against the system has to be part of the solution—in fact a large part, given the limits and pitfalls of electoral politics. This is clear enough—at least to indigenous leaders like Humberto Cholanga, ECUARUNARI's current leader, who have a clear class perspective on the 'indigenous question.' But another part of the solution is to create conditions that will facilitate the birth of the new. We know that the process will be fraught with attendant difficulties and will need the services and support of a midwife, to speak metaphorically about the dynamics involved. A close look at these dynamics will not give us the answers to these questions. But it might very well put the protagonists of the class struggle in Ecuador on the right path.

5

The Politics of Adjustment, Reform and Revolution in Bolivia

Bolivia's political dynamics since 1985, the year in which the government turned toward neoliberalism, warrant a closer look and further study for at least two reasons. One is that the country has served as a major experimental laboratory to test both the Washington Consensus on 'correct' macroeconomic policy and the new policy agenda of the 1990s—an agenda that includes a new social policy to give structural adjustment a more human face, a local form of sustainable development, and a new governance regime. An examination of the economics and politics of adjustment in Bolivia is revealing of the agenda pursued over the past two decades by virtually every government in the region under the aegis of global capital. An analysis of the political dynamics of this agenda is one objective of this chapter. But there is another reason for reviewing political developments in Bolivia since 1985. No other country in the region provides as clear a lesson about the political dynamics involved in the relationship of the state to the social movements. The dynamics of this relationship, and its theoretical and political implications, are the central concerns of this chapter.

A HISTORY OF EXPLOITATION, OPPRESSION, AND REBELLION

Bolivia is one of the poorest countries in South America. To understand this poverty and the conditions that continue to reproduce it is necessary to some extent to delve into the past, both the recent past as of the 1952 revolution that generated a new dynamic in Bolivian society and politics, and the more distant past of European conquest, colonial rule, and postcolonial developments. In this history we can identify three distinct forms of society and politics, each evolving into a world of experience that is at once distinct and separate, but yet articulated and superimposed one upon the other (García Linera, 2004). The first is based on a capitalist mode of production and takes the ideal-typical form of a class-divided society, a market economy, modern industry based on bourgeois relations of production, a capitalist state, and a culture of possessive individualism and competitiveness. Another 'civilization regime,' to use Garcia Linera's terminology, or 'social formation,' to use a term fraught with fewer difficulties, is based on a simple (or domestic) mode of production, articulated with the dominant capitalist mode but with its own superstructure of relations among small landholders, artisans, and family-based (non-communtarian) peasant producers (*parcelarios*)—relations that are projected symbolically and represented politically in a culture of autonomy and communitarian democracy (García Linera, 2004). A third form of society or civilization is rooted in a communal mode of production and characterized by institutions that generate a spirit of community and relations of solidarity and reciprocity among individuals who prioritize 'the community' and sharing of productive resources over private property.[1] In the Bolivian context this sociocultural regime is embodied in the community, or commune, that defines the way of life, religious beliefs, values, and traditional form of political authority shared by diverse indigenous groups—predominantly Aymara and Quechua.[2] The indigenous population of Amazonia represents a fourth form of civilization and society that is communal in form but smaller in scale, rooted in a hunting and gathering type of economy and characterized by the absence of a state.

As for the 1952 'revolution,' which came upon more than 125 years of postcolonial rule, it was not really a 'revolution'—not in the sense of Cuba, for example. But it did bring into motion dynamics that brought the largely indigenous working population into a slow process of economic and social development, albeit a system dominated by a small governing and non-governing elite. Among other changes the 'revolution,' led by the MNR but enacted by the tin miners and the peasantry, and buttressed by militant peasant leagues and working-class unions, gave rise to political developments that would culminate in a short-lived Democratic and Popular Unity government organized on the basis of workplace assemblies and defended by armed worker and peasant militias. But the government formed under these political and economic conditions was assailed by forces of political reaction and economic instability generated by these forces. Neither this government, nor the authoritarian military regimes that preceded and followed it, were able to secure economic development or political order. What we see is an extended period of political and economic crisis, reflected in spiralling rates of inflation and a succession of short-lived governments.

The political dynamics unleashed with this development are difficult to summarise briefly. But with a focus on their contemporary relevance we note the following:

1.   Although no US economic interests were directly threatened by the revolutionary regime, Washington was nevertheless alarmed. The Eisenhower administration decided that rather than confront the revolution it would disarm it. By providing substantial food aid, development assistance, and subsidies to the beleaguered tin industry Washington gained influence with the moderate center-left of the governing party, the MNR. Gaining the upper hand, these moderates tempered their anti-American rhetoric and distanced themselves from labor radicals and communists. The hand of the moderates was further strengthened by the tepid land reform measures, which converted the peasantry into a conservative force, something that it has since learned to regret.

2.   The Central Obrera Boliviana (COB), formed on April 17, 1952 with the organization of over 25,000 miners under the leadership of the Juan Lechín, became the principal political instrument of the labor movement, playing this role throughout the subsequent 50 years of political developments. In 1985, with the installation of a neoliberal MNR regime, the hitherto powerful labor movement was seriously weakened through the effects, first of all, of the closure of the mines and the dismissal of most of the miners, the backbone of the labor movement and the dominant sector within the COB. A series of subsequent policy measures decimated the ranks and the leadership of the COB, significantly reducing its role on the national sociopolitical scene (Salvatierra, 1999: 3). The dynamics of this process included the creation of a 'free' labor market, a new mode of labor regulation resulting in greater flexibility, negotiations of collective agreements at the level of the firm, the disappearance of social benefits, a generalized substantial decline in the value of wages, increasing informalization (up to 64.7 per cent of the labor force in 2002), and widespread fear of further dismissals.[3]

3.   Notwithstanding a land reform program that redistributed land to some 256,000 peasant families and the unionization of the *altiplano* peasantry,[4] the revolution did not substantially affect the marginality of the indigenous population and their communities in the highlands and the Amazonian region. Many of these communities were integrated

into Bolivian society and the state for the first time under the reforms instituted by De Lozada's first administration (1993–97), which converted community-based indigenous organizations into OTBs (Organizaciones Territoriales de Base—territorial community-based organizations) with the rights of representation and participation in a series of municipal or local development plans.

4.   The military–peasant pact engineered by the MNR government as a weapon to be used against the powerful mining unions began to unravel in 1971 with the emergence of the *Kataristas*, a movement of *Aymara* peasants who gained control of the peasant confederation and introduced, for the first time, an explicitly ethnic dimension to the class struggle (Albó, 2002: 76–7). The National Confederation of Peasant Workers of Bolivia (CNTCB), formed with MNR support, was restructured and became the more autonomous and politically radical Confederation of Peasant Workers' Unions of Bolivia (CSUTCB), newly affiliated with a rejuvenated COB with the declared purpose of defending their land against the predations of the landowning oligarchy and the bourgeoisie.[5]

5.   In 1982, in the context of a general strike and widespread mobilization that ended Bolivia's period of military rule, the indigenous communities of Bolivia's eastern lowlands also formed a national organization to represent their interests and advance their struggle—the Indigenous Confederation of Eastern Bolivia (CIDOB). Unlike the western highland counterparts the lowland *indígenas* (indigenous) did not see themselves as *campesinos* (peasants) or *trabajadores* (workers). They were not peasants in fact, and they defined their struggle not in class terms but in terms of a demand for territory rather than land, autonomy, and the recuperation of ethnic culture rather than economic and political concessions from the state. In these terms, in 1994, under De Lozada the CIDOB won important changes to Bolivia's political constitution in the recognition of indigenous peoples as well as their territorial and communal rights to natural resources, and their traditional norms and forms of leadership. These concessions reflected in part De Lozada's agenda of co-opting the indigenous movement, to prevent its linking up with the peasant movement. However, they can also be seen as the government's response to a highly politicized mobilization, in 1990, in the form of the 'March for Territory and Dignity' in which representatives of the indigenous communities marched for 70 days

from the eastern lowlands toward a symbolic encounter with their *Aymaran* counterparts in La Paz.

6.   The neoliberal model was installed in 1985[6] as a means of activating the national economy—to create the conditions for economic recovery and stable growth. However, there has been next to no economic growth despite heightened levels of capital formation and economic development effort. Growth over the subsequent reform period has been virtually non-existent, reflected in widespread conditions of social exclusion and persistent poverty (over 60 per cent of the population).

7.   Instead of economic growth neoliberalism in Bolivia brought about increasing levels and diverse forms of organization within a burgeoning 'civil society' (peasant organizations, unions, labor centrals, civic committees, and neighborhood associations) and an intensification of the class struggle and associated political conflict and social movements. The initial effect of the neoliberal model was a serious disarticulation of the labor movement and the ending of what has been described as 'the golden age' of Bolivia's peasant movement which saw widespread mobilization, the strengthening of *campesino* political participation, a heightening of ethnic identity, and the emergence of a new political project—the construction of a multi-ethnic state. However, after more than a decade of low-intensity class warfare, dominated by a process of community development and a turn toward local politics, the class struggle has intensified. This process is reflected in an almost continuous process of mass mobilization since April 2000.

THE POLITICAL ECONOMY OF ADJUSTMENT, 1985–2002

Political and economic developments in the 1980s and 1990s in Bolivia, indeed in most of Latin America, responded to a perceived need within the ruling class to adjust to the dictates of a New World Order. This process was advanced on the basis of what in Latin America is termed the 'new economic model'—a model, which in terms of its theoretical propositions and policy prescriptions, is understood alternatively as 'neoliberalism,' 'market fundamentalism,' or the 'Washington Consensus.'

The construction of this model can be traced back to the experiments in policy reform conducted by the military governments in Chile,

Uruguay, and Argentina in the 1970s. As to why these regimes turned toward a neoliberal model for the construction of what one of the architects of the 'structural adjustment program,' with reference to Chile, termed 'the most sweeping program of economic reforms in history,' the question is as yet unsettled. But it is clear enough that like the subsequent counterrevolution in development theory and practice which Toye (1987) and others associated with monetarism and conservative politics and fundamental economic reform in both the US (Reagan) and the UK (Thatcher), the turn toward a new economic model in Latin America reflected a conservative reaction by property-owners in regard to both land and capital, against a protracted process of state-led reform that they perceived to be a threat to their property rights, eroding the right of property-owners to dispose freely of their assets ('freedom' in neoliberal discourse).

In the southern cone of the Americas (and later in the US and Europe) the impulse toward fundamental economic reform was essentially the same. Although couched in terms of perceived failure of the existing economic model (import substitution, state protection and developmentalism, etc.) and conditions of a widespread fiscal crisis, the neoconservative counterrevolution in Latin America, Europe, and the US was predicated on a perceived need to put an end to and reverse a process of economic and political development associated with the welfare state in the North and the developmental state in the South.

This process was advanced by use of a nationalist/developmentalist/populist economic model (see, in particular, Brazil and Mexico) which emphasized:

(1) the nationalization of industry in strategic sectors—oil and gas, mineral extraction—and setting up public enterprises in these sectors;
(2) the assumption by the state of the theoretically (and politically) defined 'functions of capital'—productive investment, entrepreneurship, business management and economic planning;
(3) protective measures in regard to national industry, insulating national enterprises from undue foreign competition;
(4) industrialization in the form of an import-substitution policy;
(5) an inward orientation of the production apparatus toward the domestic market, with measures designed to increase

the purchasing power of wages and thus maintain market demand;

(6) the regulation of product, capital and labor markets and the restriction of foreign direct investment;

(7) reform measures designed to improve the access of disadvantaged groups to society's productive resources and to increase the share of labor or wages (incomes available to workers and households) in national income; and

(8) the redistribution of market-generated incomes in the form of social programs designed to meet the basic needs of the population.

The counterrevolution can also be viewed as a strategic and political response to the system-wide crisis that put an end to the 'golden age of capitalism' in the early 1970s. Other responses included:

(1) The separation of labor-intensive operations of industrial production from capital-intensive operations, and their geographic relocation closer to available and new sources of cheap labor, creating in the process a 'new international division of labor';

(2) the development of new production technologies related to microelectronics and the computer chip, and the technological conversion of the industrial production apparatus, creating a process of 'productive transformation';

(3) the search for more flexible forms of organizing the labor process at the point of production, creating a new post-Fordist regime of accumulation and mode of regulation;

(4) on the part of the US Reserve Board take the US dollar of its fixed exchange against gold, letting the dollar float as a means of offsetting the comparative advantage of both Japan and Germany vis-à-vis exports, and creating thereby moiré favorable conditions for US exports;

(5) a direct assault of capital on the capacity of workers to organize and negotiate better working conditions and higher wages, viewed as the cause of both inflation and a crisis in productive investment, placing undue pressure on capital and leading it to be withdrawn from the production process; and

(6) globalization—cresting a global economy based on the free exchange of commodities and the free movement of capital,

freed from the constraint of government regulation and the restrictions placed on it by the state.

Whatever the precise origin of the model used in the reform process there is little question that in both Europe and Latin America it had its political roots in a conservative reaction to a process of state-led economic and social reform and an associated pattern of economic development—the gradual incorporation and increasing participation of what Stieffel and Wolfe (1994) have termed the 'hitherto excluded' sectors of society. In this reaction the basic concern and aim of the market-led reforms was to reassert the rights of property—and to release what George W. Bush, in his 2002 National Security Doctrine, dubbed the 'forces of freedom. Democracy and free enterprise,' namely private property, the private sector, and the market freed from government interference.

In effect, and not to put too fine a point on it, the neoliberal policy reform program was one of several ways of restructuring the relation of 'capital' (private owners of the means of social production) to 'labor' (producers of value in the process of capitalist production). Other such widely implemented measures included:

(1)   a direct assault on the organizational and political capacity of workers, limiting their capacity to negotiate collective agreements;
(2)   deregulation of labor markets; and
(3)   diverse mechanisms designed to compress the purchasing power of wages, leading to a reduced share of wages in national income and a larger pool of resources available from productive investment.

Arguably, the neoliberal reform program is a part of the same process—a strategy designed to activate a global process of capital accumulation and, via a reconcentration of the means of social production, generating the financial resources needed for productive investment.

THE ECONOMICS OF BOLIVIA'S ADJUSTMENT

Regardless of its origins or the motivations behind neoliberalism, Bolivia represents an important part of its construction, an important thread in the weaving of the Washington Consensus. The experiments in neoliberal reform constructed under the military regimes of Chile, Argentina, and Uruguay in the 1970s can be viewed as the first step in the practice of neoliberalism, the implementation of a doctrine constructed by von Hayek and others at the University of Chicago. The second step involved the turn toward neoconservative politics in the US and the UK, while the alignment of World Bank and IMF policies of macroeconomic stabilization and structural adjustment can be viewed as the third major development in the direction of neoliberalism. The policy reforms implemented in Bolivia in 1985, on the basis of Presidential Decree 20160, can be viewed as the fourth major step on this path, a path paved by concerted political efforts to create the ideological and political conditions for the neoliberal model. These conditions not only entailed the use of external debts as a mechanism for policy leverage but a major ideological offensive, waged on various fronts, and diverse efforts to consummate the newly proposed marriage between democracy and capitalism (Dominguez and Lowenthal, 1996).

In 1985, under conditions of an external debt of close to $4 billion and runaway inflation that reached levels unheard of even in inflation-prone Latin America (peaking at 25,000 per cent—2,000 per cent from 1983 to 1984) Bolivia's MNR government initiated a radical program of policy reforms that included (1) a sharp currency devaluation, followed by the adoption of a more 'realistic' rate of exchange; (2) measures to liberalize the entry of foreign goods and the movement of capital in the form of direct foreign investment; (3) the elimination of government regulations and subsidies in regard to product and capital markets; (4) closure of the mines and a drastic reduction of the public sector with regard to employment and social programs; and (5) a significant reduction of government expenditures and both public and private consumption. These and other policy measures were implemented so swiftly and extensively that the international financial community would deem Bolivia as one of the most successful economic adjustment programs in the postwar era.

Chile's pioneering program of policy reforms had been engineered by the 'Chicago Boys'—technocrats trained in the ideas of von Hyek, Friedman, etc.). In Bolivia's case, the 'technocrats' (economists who view economic development as a technical rather than a political issue) responsible for the reform program were largely trained at the Harvard University, advised by one of its economists Jeffrey Sachs, one of several 'Harvard Boys.'

The initial shock reform program did indeed contain inflation. But a sustainable process of economic reactivation would require a broader program of structural reforms. To this end, the economic team assembled by the government, headed by Sánchez de la Lozada, Minister of Planning at the time ('*Goni*' or '*el gringo*' as he is unaffectionately called in Bolivia) entered into a series of high-level meetings with officials from the 'international financial community' (World Bank, IDB, etc.), the UNDP and representatives of the most important overseas development associations' (USAID, etc.) operating in Bolivia. These meetings extended from 1986 to 1992, months before *Goni* assumed the presidency.

Such meetings tend to be held behind closed doors and in secret. However, in this case, we have a revealing account of their proceedings by Denmark's representative of 'international cooperation for development.' In this account there were three major 'strategic considerations' used to establish for the government its reform orientation, a fundamental legal and administrative institutionality, and specific reform measures: (1) *productivity competitiveness* (how to improve the productivity of Bolivia's major economic (i.e. business) enterprises and ensure their ability to compete in the world market); (2) *social integration equity* (how to broaden the social base of national production, improving access to means of production of diverse groups of producers beyond the small stratum of large well-capitalized enterprises privileged by, and benefiting from, neoliberal policies); and (3) *state action governability* (how to ensure political order with as little government as possible, i.e. via the strengthening of civil society and participation in public policy).

Economic and social development in the late 1980s and early 1990s was essentially state-led, that is, initiated and orchestrated by the government. At issue in this 'development'—or whatever transpired in the 1990s—was a series of reforms introduced through legislation or by administrative fiat (executive decree) as a function of a model ('sustainable human development') designed to the purpose (basically along the lines of the UNDPs' conception of 'human development') and used in the construction of De Lozada's 'Government Plan for Action' (1993–96). This Plan was preceded by a 'social strategy' to support a 'new social policy' (a poverty-targeted Social Emergency Fund) and followed by a Plan for Action (1997–2002), which, in line with a formulation by ECLAC (1990), for the first time defined the principle of 'equity' as a fundamental pillar of government policy.

In most interpretations of the policy measures introduced by the government—and these measures are widely perceived as a model for other countries—economic development in the 1990s was neoliberal in form. This is to say, it was based on the Washington Consensus on correct policy (Williamson, 1990). However, a close examination of the policies actually implemented suggest that they derive not so much from a neoliberal model as from a hybrid, a model that might better be dubbed 'social liberalism.' As for the structure of this model it is based on a 'new policy agenda' that reflected the recognition that neoliberalism is politically unsustainable and economically 'dysfunctional.'

The basic elements of this new policy agenda are:

(1) a neoliberal program of macroeconomic policy measures, including privatization, agricultural modernization, and labor reform;

(2) a 'new social policy' supported by a 'social investment fund' targeted at the poor;

(3) specific social programs (policies related to health, education and employment) designed to protect the most vulnerable social groups from the brunt of the high 'transitional' social costs of structural adjustment—and to provide a 'human face' to the overall process; and

(4) a policy of administrative decentralization and popular participation designed to establish the juridical-administrative framework for a process of participatory development and conditions of 'democratic governance.'

These policy directives were in large part implemented from 1994 to 1997. One of the final steps in this program of structural adjustment and reform, achieved through the Law of Capitalization, was the privatization of the remaining five strategic state enterprises: the national mining company, the mainstay of the Bolivian economy, and the oil, gas, airline, railway, and telephone companies. As a result, Bolivians no longer own or control their mineral or hydrocarbon resources, or their transportation and communication systems. And tens of thousands of workers have been laid off. However, on his return to power in 2002 and efforts to put the finishing touches on his policy agenda De Lozada experienced a setback in the form of massive mobilizations against his plan to sell off to outside interests the country's reserves of natural gas. A similar plan by the government in

2000 to privatize and sell off another of the country's most important strategic resources (water) resulted in a massive mobilization that forced the government to reverse its policy and drop its plan. The political dynamics of this process—the October uprising and the gas war of 2003—will be reviewed below. They include the ousting from office of De la Lozada himself, Washington's 'best friend' in Latin America—although, it has to be said, Lucio Guttiérez has disputed this (dis)honor.

THE POLITICS OF ADJUSTMENT AND REBELLION, 1993–2003

In securing the political conditions for its reform agenda the Bolivian state engaged in a series of political strategies for managing its relations with the social movements and other manifest organized forms of political opposition. These strategies can be placed into four main categories:

(1) *electoral politics*—incorporating oppositional forces and political projects into the 'system,' that is, the mechanism of electoral politics and the organization in the form of a system of political parties, each vying for state power and sharing it as required, and a number of 'oppositional' parties to consume the political energy generated in the process of political opposition;

(2) *local development projects*—providing organizations in the popular sector a clear option to social movements and their tactics of direct action and mobilizations in the form of alternative development projects;

(3) *promoting*, through the Laws of Popular Participation and Administrative Decentralization, *democracy and good governance*—providing thereby the political stability for economic globalization;

(4) *co-opting and accommodating the leadership* of the social organizations and movements to the economics of free markets and the politics of democratic elections—to the virtues of democracy and capitalism;

(5) where necessary, that is, when and where co-optation has failed, *intimidation and tactics of repression*, including police or military action (marshalling of the security apparatus) against the social movements.

2000—From April to September: Popular Rebellion and the Water War

The April 2000 'water war' of Cochabamba was a major watershed in the history of the popular movement, representing the first serious reversal in the Bolivian government's agenda in regard to sovereign control over the country's natural resources. Bolivia is also rich in resources such as oil and gas but none of these is as critical to the livelihoods of all Bolivians as water. And it was the government's intention to alienate the last vestige of public control over this resource that sparked a mobilization that compelled the government to reverse its decision to concede the legal right of Bechtel, the ultimate owner of the privatized '*Aguas del Tunari*,' to market Cochabamba's reservoir of water without regard to the needs and rights of Bolivians, particularly those in the immediate locality. The second battle in this war—a legal suit leveled against the government by Bechtel to compensate the company $25 million for its loss of anticipated revenues[7]—is ongoing, but the first skirmish in this war was decidedly won by 'the people,' a mass mobilization that brought together diverse sectors of the popular movement into a 'Coordinating Body for the Defence of Water and Life.' Other demands of the movement included the expulsion of *Aguas del Tunari*, a multinational company registered in the Netherlands and with divisions in the UK and the US, and ultimately owned by Bechtel, one of the largest resource enterprises in the world; the derogation of SD2029, which legalized the privatization of water, and the prevention of the export of water by *Aguas del Tunari* by way of Chile, a venture entailing investments of $80 million and anticipated revenues of $1.2 billion, almost all of which would accrue to a multinational company.

The popular insurrections of April and September were connected to, and to some extent gave rise to, diverse political currents or tendencies within the popular movement. Five of these currents have played a particularly important role in the trajectory of the movement since 2002:

(1) a current headed by Felipe Quispe Huanca, the *Aymara Maliku* (maximum authority), leader of the CSUTCB and founder of the *Pachakuti* indigenous movement (MIP). Like MAS, the MIP was a political party rather than a movement, and like the movement led by the historic *Tupaj Katari*, it was oriented toward Aymara nationalism rather than class struggle, seeking self-determination of the Aymara nation with territorial control over natural resources (from a peasant perspective), and, in

this context, set against colonialism in its historic and current forms as well as imperialism (the *K'ara* bourgeoisie and *K'aras* in general).[8] The aim of this anti-imperialist struggle is restoration of *Q'ullasuyu*—the proto-socialist community of the original *ayullu*.[9]

(2)   An intellectual and political current oriented toward *social* rather than political development—an alternative form of local and autonomous NGO-assisted development protagonised 'from below.'[10] With Alvaro García Linera, a MAS-linked intellectual as its chief ideologue, and the protagonism of Oscar Olivera, an indigenous leader (close to Evo Morales) who has disputed Quispe's control of CSUTCB, this current played an important role in the 2000 water war of Cochabamba (via the Coalition in Defense of Water and Life), notwithstanding an opposition to class struggle and political power. In the aftermath of the revolutionary crisis of February 2003 García Linera openly supported the 'social pact' sought at the time by De Lozado with the social movements. Unlike García Lineres and Filomón Escobar, another MAS theoretician who has taken more of a class line rather than García Lineres's position on autonomous local development, Evo Morales, in his role as leader of the Chapare confederation of coca-producing peasants (*cocaleros*) at the time rejected De Lozada's overtures. Like Angel Durán, leader of the Movement of the Landless (MST),[11] he did not reject such overtures from Mesa.

(3)   A labor movement led by the *Central Obrera Boliviana* (COB). Set up in the context of the 1952 Revolution the COB has played a central role in the labor movement ever since, notwithstanding its politics of negotiation and pacts signed with the government of the day. But in the context of the mobilizations that have dominated Bolivian politics since 2002, as the leading organization in a severely debilitated labor movement COB has played a decidedly secondary role to the indigenous movement, particularly that part of it located in (and self-identified as) the peasantry. Neither COB nor MAS played a major role in the Cochabamba water war or the gas war triggered by De Lozada.[12] Even in October 2003, in the throes of a quasi-revolution, the efforts of COB to galvanize collectively the working class, through the mechanism of a general strike, its major political weapon, were less than successful. And in the immediate wake of the twelve days (8–19 October) that 'shook Bolivia' to its political

roots COB, like MAS, was quick to demobilize the insurgency, opting for a 'tactical retreat' from the class struggle and—like several other organizations (MAS, MST)—'critical support' of the new regime.

(4) An anti-neoliberal bloc constituted by a sector of the organized labor, headed by the *fabriles* of El Paz, independent from both 'the system' and the COB. This bloc opposed the dominant political line of negotiations and dialogue with the government pursued by the COB over the years.

(5) The political current represented by Evo Morales in the leadership of the coca-producing peasants in the eastern lowlands of Chapare, defines the politics of an important sector of indigenous peasants. Unlike the sector represented and led by Quispe these indigenous peasants are not unionized and were organized first and foremost to protect their production of the coca, which, as Evo Morales pointed out in an interview with *Punto Final*, represents not only the livelihood of tens of thousands of indigenous peasant families but their sense of collective identity—the 'backbone of the *quechua-aymara* culture,' and as such, 'something with considerable economic and social value' (Morales, 2003: 16). In the context of political developments in 2002 this political current became part of a larger popular movement, the defence of coca assuming the same significance as oil, gas and water, i.e. as an issue of national unity in a popular struggle against neoliberalism.

Pachakuti (MIP)

The decision of Quispe to create the MIP in the context of the popular uprisings in 2000 generated a crisis within the indigenous movement, the leadership of which suspended Quispe and several others for 'treason and for having violated the organic statutes of CSUTCB' (*Presencia*, la Paz, November 26, 2000; *Pulso*, November 27, 2000). In fact, delegates from eight government departments approved a resolution to 'prevent the return of *Maliku*.' Even so, Quispe remains the most important leader of the insurgent forces of El Alto, a predominantly Aymara city[13] next to la Paz, and in this political context was demonized by the press throughout October. In fact the reactionary Catholic paper, *Presencia*, recognized the MIP's solid base in the indigenous sector and that 'stripped of sectarianism and ... exclusions, myths and illusions, the movement could very well succeed in its aims' (November 17, 2000).

2002—A STORY OF TWO MOBILIZATIONS AND AN ELECTION

The year 2002 opened with a series of economically damaging and paralyzing highway blockades organized by the federation of ex-mining coca-producing peasants led by Evo Morales. The blockade strategy was announced over a month earlier in response to the refusal of De Lozado to respond to the demands of the coca-producing 'peasants' (*cocaleros*) of Chapare and the Yungas, or even to negotiate, a halt to the government's US-mandated coca eradication campaign, a policy of devastating consequence to the *cocaleros* in the context of a totally ineffective alternative crop development program. More than a week into a heated conflict between the government and the *cocaleros* in Chapare, Felipe Quispe, '*El Maliku*,' the dominant indigenous leader in one of the most heavily populated urban centers in the country, joined the struggle. After a series of violent confrontations the government finally relented from its hard line and signed an agreement with Morales that ended the social conflict, or at least inaugurated a temporary truce—a truce that the government sought, and more or less achieved, with the other sectors of the popular movement. In the agreement signed with Evo Morales, the government backed away from its decrees governing the eradication and commercialization of the coca, with the intent of redrafting another decree over the next three months. Thus De Lozada bought himself some time, but April and September of that year saw other major irruptions of social protest and mobilisation of direct action across the country.

The next critical conjuncture involved the mid-year presidential and parliamentary national elections. After a spell out of office De Lozada was back as the MNR's presidential candidate, this time running with Carlos Mesa, a former journalist without de Lozada's ties to the World Bank and the US but with excellent credentials as an MNR militant. The most dynamic and relatively new factor in these elections was the participation of MAS and its presidential candidate—Evo Morales. His participation in these elections at the time raised serious concerns within the ruling class and its political representatives in '*el sistema*' (a cluster of rightwing parties (MNR, UDN, UCS, NFR).[14] These concerns, fed by polls that put Evo Morales in third place and that predicted MAS would elect from 20 to 28 deputies, were that this intrusion of the indigenous factor (and the popular movement) into what had been their exclusive preserve would directly threaten the government's agenda and the economic interests behind it. To that date, the parties of the system had always disputed and shared power among themselves, with the tacit COB or—but now open—support from the US.[15]

In the event (30 June) the MAS challenge was enough to push the leading government party (MNR) into an electoral and government alliance with not only the NFR and MBL, its traditional coalition partners but with the MIR, a party ostensibly on the left (and in opposition to the government party) all too ready to join any government coalition in exchange for a share of power.[16] This coalition managed to squeeze out an electoral victory for De Lozada Mesa over Morales/MAS—22.45 to 20.94 per cent.[17]

Notwithstanding the better than expected results for MAS and the machinations of a rightwing alliance, some observers interpreted the results as evidence that a quarter of the population supported the government in its orientation toward a free market economy and the denationalization of Bolivia's vast reservoir of natural and productive resources (*30 Días*, June 2002). But the elections had a broader significance. For one thing, the elections masked a historic turning point. Against a campaign headed a former president and millionaire mining executive, a social movement of peasants and indigenous peoples came within an inch (1.4 per cent) of wresting control of the political power that had eluded them for 500 years. For another, the resurrection of the left, on the base of a dynamic indigenous movement, and in the multiple form of a social movement and electoral politics, unmasked the democratic façade that the ruling class had for so long thrown over its machinations for state power and the belief that it had an inherent right to rule as well as to protect their class interests. For another, the elections represented a change in the dominant strategy that the left would use to seek state power, shifting from a reliance on social movements and mass mobilization to electoral politics within the institutional framework of the existing liberal democratic system.

More generally, the turn toward electoral and parliamentary politics also brought about a realignment of state power. In this realignment the economic power in the country as always was ensconced in the executive, which implemented its reform program as a series of 'supreme presidential' decrees, but in the legislature where MAS was a force that could not be denied or ignored, there was a significant reconfiguration of political power. Even so, although it was not this force that would bring down the government, neither De Lozado then (before October) nor Mesa now (after October) was able to govern without the 'functional opposition' (that is, critical support) of Morales and MAS. It is significant, albeit not surprising given the circumstances, that this support was forthcoming.

Electoral and parliamentary politics, however, did not mean the end of the popular movement and the dynamics of mass mobilization. On the contrary, the social movements in Bolivia continue to exhibit significant dynamism in spite of diverse efforts to demobilize them. A case in point is presented by a major pre-election mobilization in the form of a 30-day march on La Paz, organized by a social movement formed by diverse indigenous lowland and some highland groups under the leadership of *Bienvenido Zaco* (CIDOB—*the Confederación de Pueblos Indígenas del Oriente*) and *Faustima Zegarra* (Conamac). These and other popular organizations[18] formed a movement in support of a new constitutional assembly that would bring about a multi-ethnic and pluri-national state that is more inclusive and participatory, i.e. democratic. In the event of this march on La Paz, in the lead up to the June 30 elections, there was no agreement on tactics and strategy vis-à-vis the government, which was intent on ensuring that this mobilization would not interfere with the elections. The groups associated with CIDOB and led by Zaco agreed to meet and negotiate with the multi-party government commission. However, the largest column in the march, from Cochabamba and headed by Zegarra, rejected the government's proposal to meet with the commission, insisting that any such meeting would have to be in the open and negotiations conducted at the site of the march (*30 Días*, June 2002, p. 19).

The results of these negotiations were ambiguous. On the one hand, the marchers and the government, via its inter-party commission, came to an agreement that a constituent assembly would be placed on the government's political agenda, and the proposal to establish such an assembly would be taken up by MIR—and, in the wake of October, by the Mesa regime. But this would be two years down the road and it was not at all clear that the proposal for a constituent assembly would have the outcome initially desired by the movement for a pluri-national, more participatory, and socially inclusive state. However, on the government's side it managed to defuse the conflict surrounding the issue of a constituent assembly and to create conditions that allowed the 2002 elections to take place.

Not all mobilizations have had such dubious outcomes. In fact, a number, particularly those that have held to a class line of struggle, have been successful in achieving their demands. Take the case of workers in the Huanuni mine near Oruro. In the same pre-electoral context of 2002 mobilization in the form of a blockade of highways resulted in the government backing down and reversing the *de facto* and legal privatization of the mine (*30 Días*, June 2002, p. 25). And these workers were not alone. Their mobilization was very issue- and workplace-specific, involving not only the miners themselves but diverse sectors of the popular movement. Subsequently, however, a series of mass mobilizations managed to halt government efforts to extend or deepen its agenda. Notable cases of successful mobilizations involve what became known as the 'water war' in Cochabamba and, three years later, the 'gas war.' In the case of water the government had to do a u-turn, but in the case of gas (whether or not to sell off its gas reserves, the second largest in Latin America, or restatify and nationalize it) the government took the route of holding a referendum. The referendum, set initially for July 18 2004 but postponed until August, was politically contentious in that the dominant position of the popular movement, held by Quispe and Solares among others, is that 'the people have already spoken,' i.e. they set against the government's plan to privatize the exploitation and sale of gas, and to export it via Chile. In fact, opinion polls indicated that over 60 per cent of Bolivians were against the government on this issue.

2003—THE MAKINGS OF A REVOLUTIONARY SITUATION

Many observers trace the origins of the October uprising to policies introduced by the government from August 2000, just a month into De Lozado's second administration, to February 2003, when De Lozado followed its *'perdonazo'* (the pardoning of $180 million in debts held by the largest corporations)[19] with an *'impuestazo'* (tax hike) in which the government sought to raise an additional $90 million in taxes from the working class. Although the government managed to gain some breathing space and several months of relative calm, after an initial popular rebellion of pensioners in January and an outbreak of violent clashes with government forces in February over the *'impuestazo'* which resulted in 35 deaths and 250 seriously injured, a revolutionary ferment of opposition and resistance was brought to boiling point in October by the government's proposal to concede the right to export its massive strategic reserves of natural gas to a multinational, US-owned consortium. Under conditions of this project the political impasse in the workers and broader popular movement was broken and the government confronted its worst political crisis for many years, a crisis that forced De Lozado into exile in the US.

In the run-up to this crisis, the government sought to consolidate a political base for its 'mandate' by bringing into the government the NFR, its primary opposition within the system and hitherto opponent of its privatization agenda and the proposal to concede all rights to export natural gas to Pacific LNG, a consortium of YPF, British Gas, and Pan American Energy. With this political move, costing the government three ministries and 15 vice-ministries, as well as the strategically important prefecture of La Paz, the government was supported by a mega-coalition of MNR, NFR, MIR, and UCS. Having thus consolidated its political base (in July 2003) the government embarked on its political agenda, provoking widespread opposition and resistance, insurgent forces which came together in the October uprising—in what to some had the makings of a social revolution. In the wake of the February rebellion against the *impuestazo*, and in an effort to prevent the threat of further social conflict, the government legislated a new 'code of citizen security' which prohibited road and highway blockades, a major and increasing tactic of popular struggle. At the same time it moved to bring the insurgents to the negotiation table, taking a carrot-and-stick approach and mobilizing its repressive apparatus, killing on average five workers a day in the eleven months of De Lozada's second term in office. Under these conditions in August the government also attempted to reintroduce the *impuestazo* but in a different form. Notwithstanding the likelihood of provoking another rebellion, the IMF applauded the government's 'bravery' in seeking to 'recover fiscal balance' and 'macroeconomic equilibrium' (i.e. meet the government's debt obligations)—at its behest, the IMF might have added—by 'moderniz[ing] the tax system.'

For many who joined the popular uprising what was evident was its apparent spontaneity, that is, the relative lack of political organization or mobilization, which, it appears, followed rather than preceded the initial outbreak of conflict. In the successive waves of this insurgency the whole country was paralyzed by the mobilizations that followed, including a general strike of peasants, workers, students, and a multitude of other popular organizations, but these actions were not mobilized on the back of a clearly defined project, one of several reasons why this wave of revolutionary ferment would eventually crash on the rocks of Mesa's 'government of transition.'

At the base of the insurgency and associated mobilizations, was a proletariat of *Aymaran* peasants in El Alto, an *Aymaran* city next to the capital. By various dispassionate accounts (discounting the claims of some individuals and organizations) no political organization could claim hegemony over, or assert control of, this historically important wave of insurgency, not even Felipe Quispe *Maliku*, the charismatic leader of the most important organization of insurgent forces, the *Aymaran* population of El Alto, as well as CSUTCB and the MIP.

On October 12, 26 *alteños* (residents of El Alto), were massacred by government forces dispatched to quell the several days-old popular uprising of the proletariat.[20] Over the next week virtually all of La Paz was brought to a standstill by a crippling general strike and the blockades of major highways; on October 18 thousands of *alteños* joined neighborhood groups from the steep hillsides of La Paz, miners, teachers, students, market women, butchers and bakers, truckers and taxi drivers, even elements from the middle class (intellectuals, human rights activists, professionals) mobilized by the social or nongovernmental organizations, to stage the largest rally in Bolivia's history, as many as 500,000. Later in the day, when truckloads of miners and *Aymara-Quechua* community peasants ascended El Alto on their way home, thousands of *alteños* lined the streets to cheer them on, in gratitude for the widespread solidarity in their struggle. In fact, the popular uprising in EL Alto/La Paz was supported by social movements across the country, from the *cocaleros* in the eastern Chapare lowlands and the *Quechua-Aymara* community peasants from the southern highlands and valleys of Potosi and Sucre, to the miners from Huanuri (Oruro) and the multi-ethnic, cross-class civic movements that shut down Cochabamba, Sucre, Potosi and Oruro on October 14 and 15.

Most political analysts identified Lozada's intention to sell off and export Bolivia's vast gas reserves as the most explosive issue in this revolutionary situation; and indeed this agenda is one of three critical issues placed on the referendum proposed as a politically calming measure by Mesa. But according to some observers and unionists of La Paz (*Informe CITE/CODLP/265/04*) this was only partly true. Although every organized group in each sector of the popular movement raised the banner of the gas issue and generally shared opposition to the government's neoliberal agenda, they each joined the struggle for their own specific reasons.

By most accounts the revolutionary situation was rooted in, if not precipitated by, the actions of rebellious peasants and mobilized indigenous communities. But as the epicenter shifted to the cities, the popular struggle acquired a distinctly proletarian character, dominated by the indigenous sub-proletariat of El Alto, over 10,000 of whom hit the streets of La Paz, marching on government buildings and government party headquarters, and blockades. COB declared a general strike and the proletariat of La Paz and El Alto was soon joined by workers in the other major cities—Cochabamba, Santa Cruz, Oruro, Potosi—who converged on government buildings and official party headquarters, with the support of diverse neighborhood groups and sectors of the middle class that joined the struggle out of solidarity. Whatever the trigger of the uprising, and it seems to have been the government's attempt to move on its project to sell off its remaining, albeit limited, sovereignty over what is widely regarded as its most critical or strategic natural resource, the government confronted what many analysts came to regard as a quasi-revolutionary situation. Rebellious proletarianized 'peasants,' radicalized indigenous communities and organizations, and a broad swathe of the working class dominated the uprising. But what converted this explosive wave of insurgency into a revolutionary situation was the support, if limited participation, of sections of the middle class, a large sector of whom was thoroughly disaffected by the government's neoliberal agenda. Even elements of the dominant and ruling bourgeoisie were divided by policies that benefited only the banks and the privatized and denationalized sector of big capital. For ten to twelve days, a combination of these and other insurgent forces shook Bolivia to its political foundations, forcing De Lozada into US exile.

The revolution, however, did not materialize. Lacking a clearly defined political project, the diverse social forces combined in struggle fragmented, creating space and conditions for a 'constitutional solution' to the crisis and more peaceful forms of middle-class protest (hunger strike). Nevertheless, the confrontation between the insurgents and the state left in its wake 80 dead and over 200 wounded. The MAS meanwhile, as the major party of opposition to the coalition government, supported NFR's call for a constitutional solution to the crisis and proposed a constituent assembly, a position supported by the COB. Thus did the political left help ensure that the state would not fall into the hands of the insurgents or social movements.

In the particular conjuncture of October 17, having abandoned De Lozada's ship of state before it ran aground, Carlos Mesa, described by the *Financial Times* (October 20 2003, p. 5) as 'a man of the people' (versus De Lozada who 'never listened to anyone but a tiny group'), positioned himself to assume state power, and, in the process, to restore 'order.' He did so under the watchful eyes of the country's armed forces and with the support of the US Embassy, a besieged parliament, and the 'tactical withdrawal' of COB. As for MAS, it was nowhere to be seen. In the heat of the popular uprising, its leaders, including Evo Morales, like the leadership of COB, seemed to have beaten a strategic retreat from the class struggle. In this conjuncture Mesa was installed as the head of a 'government of . . . transition,' an interim constitutional regime committed to a non-violent resolution to the problems that brought about the social and political explosion of October. The referendum on the issue of gas and oil (whether or not to restatify the exploitation and production of these strategic resources) was a critical component of Mesa's 'solution,' a solution opposed by the more radical elements of the popular movement (especially the forces led by Quispe in El Alto but also the COB leadership as well as the rank and file of working-class organizations in the 'anti-neoliberal bloc') but accepted by the political class engaged in an electoral project. This included MAS, which has remained steadfast in its support of the Mesa regime on this and related issues throughout the aftermath of October.

At the time, Filemón Escobar, one of MAS's key strategists, continued to push the electoral line (municipal elections in 2004, presidential elections of 2007) over that of mass mobilization, a line eventually supported by Evo Morales, the MAS presidential candidate. Escobar had already pushed this electoral line after the February mobilizations when he met union leaders and openly declared the need to sustain the 'governability' of the De Lozada regime until the 2007 elections, following the strategy and example of the Workers Party in Brazil. By June 2004, however, he began to sing a different tune, and, in the heat of an internal debate with Morales, pushed for MAS to be subordinated to the social movement in the form of COB.

In the immediate aftermath of the October uprising leaders of diverse social organizations and movements, each according to their own ideological persuasions and interests, took different paths. Some, like Morales (MAS) and Durán (MST), supported a political truce with Mesa's transition government. Others, like Quispe and Solares, took a more radical line, although willing to concede Mesa some time. The government, in this situation, pursued the well-tried tactic of offering positions to community or movement leaders. Some (by no means all) well-known NGO researchers and development practitioners who are affiliated or work with the social movements were thus silenced or compromised. Some organization or community leaders negotiated to obtain or secure the appointment of diverse government ministries and prefectures. Thus the country witnessed a complex process that has by no means been consolidated, and the struggle for space within the structure of political power and the state apparatus continues. The efforts of these leaders of civil society to accommodate or divide the forces of resistance also continue.

THE MESA REGIME: FROM NEOLIBERALISM TO NEOLIBERALISM

The cabinet assembled by Mesa included various disciples and appointees of Jorge 'Tuto' Quiroga, political heir of the dictator Hugo Banzer and now leader of the 'renovation' and 'technocratic' fraction of Acción Democrática Nacionalista (ADN). The key ministries of Finance and Economic Development went to Javier Cuevas and Javier Nogales, both members of ADN and both well known for their connections to the World Bank. The presence of ADN members in key ministries has converted militants of this party, who still have a strong influence or presence in the union movement, into arms of the 'new officialism' and, as a result, has led to support for the government from some sectors such as the Post Office (ECOBOL), the Bolivian Mining Corporation (COMIBOL), the National Health Institute (Caja Nacional de Salud), the National Medical Association (doctors), the National Enterprise of Rural Telecommunications (SENATER), and diverse municipalities and intermediate workers centrals (COD). These and other organizations opted to support the government, as did organizations controlled by government party functionaries. An ex-functionary of this party, Roberto Barberí, and Carlos Hugo Molina, the prefect of Santa Cruz, both also with close ties to the World Bank and the IDB, became respectively Minister without Portfolio and the Minister in charge of the government's Popular Participation Program, the lynchpin of Bolivian neoliberalism.

A third critical component of Mesa's cabinet included militants and ex-militants of Movimiento Bolivia Libre (MBL), who were recruited by Mesa through NGOs with considerable influence within the social movement, neighborhood associations, and unions, including, in particular, the CUTCB (*Confederación Única de Trabajadores Campesinos de Bolivia*), which has also taken a political turn toward support of the government. This turn is not incidental to Mesa's appointment of various MBL 'personalities' to Peasant Affairs, Labor, Health, the Child Food Program (PAN) and the national Agrarian Reform Institute (INRA) as well as, most scandalously, the appointment of

Antonio Aranibar as head of the new government department of Oil and Gas Development.

MAS and MESA

In Bolivia there is a handful of systemic parties that contest state power and frequently share it, as well as a number of smaller parties, mostly on the center-left, that represent the opposition but operate within the rules of electoral politics. In the 2000 elections the Movement Toward Socialism (MAS), an amalgam of diverse oppositional forces including Morales' *cocaleros*, a movement of coca-producer indigenous peasants from the eastern lowlands of Chapare, registered as a political party and entered the electoral contest. As it turned out, the MAS candidate for president, Evo Morales, came within an inch (1 per cent of the national vote) of becoming president and defeating the eventual winner, Sánchez De Lozado, who, within a year, was forced out of office by the popular movement.

The relation between MAS and the government arguably is one of the most critical, if not decisive, factors in the current balance of political forces, representing as it does among the most important oppositional forces on the political left, within both the *cocalero* movement and the labor movement, forces that are divided as to the chosen path toward political power and vis-à-vis the Mesa regime.

In this connection, for reasons of his own (see discussion above) it would seem that Morales backed himself into a position of 'critical support' for the regime's agenda and timetable, surrendering leadership of the insurgent social movement to Quispe and other movement leaders who have been actively engaged in mobilizing the forces of resistance to the government and its policies.

The shift of MAS (and Morales) into a position of critical support for Mesa'a transition government is more than a matter of conjunctural tactics. It turns out that several 'politicos' who were brought into the Mesa government and cabinet either represent or were supported by MAS. This is the case, for example, of Donato Ayma Rojas, a union leader appointed by Mesa as Minister of Education, and Justo Seoane Parapaino, Minister without Portfolio responsible for the Affairs of Indigenous Peoples and Nationalities. Justo Seoane publicly announced that he was a candidate for MAS in the municipal elections. In addition, it seems there was a MAS connection to various social organizations represented in the government at the prefecturate level.

The 'critical' support of MAS for the Mesa regime not only reflects a calculus of political power but a strategy of maintaining parliamentary rather than mobilizational pressure on the regime as well as seeking to influence its policies 'from within.' The problem, however, is that, as in the case of education and the institutionality of the reform process, its technical and bureaucratic apparatus in control of the budget, etc. is managed by bureaucrats, most of whom were 'institutionalized' by the regimes of Banzer, Quiroga, De Lozada, and who could be viewed as World Bank technocrats. Not only do the few MAS-supported politicians in the government have to operate within an institutional and policy framework set by the World Bank, and representation of these policies in the key cabinet ministries, but they have little to no freedom of maneuver even within their limited spheres.

The question that arises is, why are Morales and MAS taking the position that they are taking? Have they fallen prey to the electoral and parliamentary politics and the reform agenda? The official line of MAS to its militants and activists is the need for a 'progressive' and 'intelligent' approach to power in the face of a 'dangerous conjuncture' between 'democracy,' i.e. the constitutional government of Mesa, and the threat of an 'oligarchic' coup machinated by the MNR, el Movimiento de Izquierda Revolucionaria (MIR), and Nueva Fuerza Republicana (NFR), the parties ousted from power in October. In this situation, MAS took the line of 'vehemently rejecting' the 'extremist position' of Jamie Solares and the COB executive who, together with Quispe, proposes mobilizing the forces of mass resistance against the government, thus playing into the hands of the coup-makers. Needless to say, the line taken by Morales and the MAS executive is very different from the revolutionary line of mass mobilization taken by Morales not that long before as leader of the *cocaleros*. In its latest national congress in Oruro the MAS executive ratified the need to contest and to gain power by 'peaceful and constitutional means,' i.e. via the ballot box, in the referendum scheduled by Mesa, the establishment of a constituent assembly and the next municipal elections in which 200 municipal governments are up for grabs. In other words, Morales has abandoned the mass struggle in favor of electoral politics, not a surprising turn of event; in fact, given the almost universal experience in similar situations elsewhere, it could probably have been predicted. Few conclusions about the dynamics of the political process are as definitive. It could almost be expressed as political law: participation in electoral politics is designed to weaken and demobilize revolutionary movements; every further step in electoral politics is a step backwards or away from the politics of mass mobilization—from the popular movement.

MAS and the Anti-Neoliberal Bloc ('el bloque')

Until the October uprising of 2003 MAS was part of the 'anti-neoliberal bloc' dominated by organizations working politically within COB such as the Department and Regional Workers Centrals, the Federation of Unionized Miners (FSTMB), the Confederation of *Fabriles*, the CUTCB, and the Federations of Pensioners, Electrical Workers, and Teacher Unions. Ad hoc organizations such as the Coordinating Committee in the Defense of Gas, in which MAS was well represented, and the *'Estado Mayor del Pueblo,'* also joined the 'anti-neoliberal bloc,' forming a pact that included the expulsion of union leaders who, over the decade, had worked closely with the neoliberal governments of the day. Jaime Solares and his executive committee was a product of this pact and took possession of COB, which began to take a more revolutionary line but was nevertheless generally supported by MAS. The anti-neoliberal bloc, which gained control of various federations of metal workers unions including the powerful La Paz Federation, had about 300 of the 900 congressional delegates of COB but as of 2003 it also captured the leadership, generally with the support of MAS delegates. After October 17, however, this conjunctural political pact between the bloc and MAS fell apart as the 'contradiction' and irreconcilable political differences between them became more and more apparent. The issue that finally divided them was Solares' proposal to 'shut down the parliament,' a position that did not sit well with MAS politicians, committed as they are to a parliamentary road to state power.

In short, the MAS leadership, including Morales himself, has taken the side of Mesa's constitutional and 'democratic' government against the revolutionary line of mass mobilization pushed by the social movements even on the leadership, notably Solares and Quispe, who might very well take a different line if not pushed in a revolutionary direction by the rank and file and social base.[21]

MAS and Felipe Quispe ('Mallku') and Pachakuti (MIP)

Felipe Quispe has played a crucially important role in bringing the *Ayamara* population in El Alto into the popular movement, but his position since October has been more difficult to situate and define. For one thing, the formation of *Movimiento Indígena Pachakuti* (MIP) was profoundly divisive. For another, he has tended to maintain a relative silence and low profile with the mass media in contrast to October when he figured prominently in media coverage. At the same time, unlike Morales, Quispe *el Mallku'* (chief) seems to have maintained his opposition to the neoliberal model and a revolutionary line of mass mobilization, joining Solares in this struggle. Nevertheless, what is less clear is his precise relationship to MAS. A number of well-known peasant leaders, well connected to Quispe, have materialized as 'advisors' to the Prefect of La Paz. When questioned on this score, the various advisors and heads of organizations under the prefecture stood by their right to 'make a living,' i.e. accept jobs or positions on offer—and based on merit.

The position taken by Quispe and his followers is understandable and politically acceptable, but it makes it more difficult to assess precisely the correlation of forces ranged between the reformist option held by MAS as an electoral apparatus and the revolutionary line of mass mobilization of the social movement. At the time of writing (May 24, 2004) CUTCB, which under Quispe more or less has held to a radical line of mass mobilization, offered Mesa's government a 30 days' truce. At the same time several public enterprises, such as EMPRELPAZ (*Empresa Rural de Electricidad de La Paz*), in practice have fallen under the control of the Federation of Peasant Federations of La Paz, one of Quispe's important support organizations. These and other such developments, together with what to some appears Quispe's apparent passivity after October, has generated criticism of his leadership among the rank and file from the position of outright opposition to the government. In this conjuncture opponents of Quispe and MIP, including MBL and MAS, attempted to capitalize politically on this criticism of, and opposition to, his leadership, in the context of their electoral campaigns in the rural areas. The electoral victory of MAS in the organization of rural teachers in La Paz is an example of this process. MAS and MIP have begun to dispute these municipal elections, in effect putting the class struggle on the backburner and the social movement on hold.

The Commune and the Anti-Neoliberal Bloc

The *Comuna* (Commune), a group led by Oscar Olivera, a labor leader of the *fabriles* in Cochabamba, is another important sector of social forces involved in the popular movement and was involved, with MAS, in the formation of the '*Estado Mayor del Pueblo*.' Prior to October the *Comuna* was well connected to the labor unions within the union-led Anti-Neoliberal Bloc and organizations dominated by MAS. But since October it appears that this connection has been severed, supporters of the *Comuna* criticizing the pro-government position of the Bloc, supporters of the latter criticizing the *Comuna* for its ultra-leftism or 'anarchism' (the proposal to close down parliament). In this conjuncture it appears that the *Comuna* does not have with it any significant unions or social organizations because most of these are involved in the process of 'building a social movement that will little by little work to undermine the existing power structures.' In this situation the *Comuna* has effectively been reduced to a few labor organizations in Cochabamba, a sector of university teachers/students in La Paz, groups of 'intellectuals,' and progressive impresarios and human rights organizations. At the same time, however, it also seems that Oscar Olivera, while escaping the spider's web of Mesa's 'distribution of positions,' together with MAS, has accepted the notion that the MNR, MIR, and NFR are conspiring in another coup, an event that would 'halt the advance of the social movements.' In effect, the *Comuna* has adopted a position somewhere between MAS and the Bloc, which, it would appear, given the turn of MAS to the right (the center to be precise) and the new position of the *Comuna*, has become somewhat weakened and disoriented. Nevertheless, Jaime Solares, a key representative of the Bloc and an 'extremist' in the eyes of MAS, in this context is pushing a revolutionary line—the need to build a movement that goes beyond the immediate demands of the organizations that make up this movement.

Although Solares does not have in mind a 'revolutionary party' along classical lines, he is clearly pushing for a revolutionary politics of mass mobilization and a rejection of electoral politics. His problem, however, is that at the moment he appears to be somewhat isolated without a mass base in support of this line. In fact, his leadership of COB is by no means uncontested or secure. For example, in Santa Cruz, Tarija, Cochabamba, and Beni a number of labor organizations within the COB have joined pro-business Civic Committees and do not recognize Solares' leadership, accusing him of being an 'extremist,' 'anti-democratic,' and a 'dictator.' They have even approached MAS to propose a 'conjunctural agreement' designed to shorten Solares' current mandate.

MAS and the Mega-Union ('megasindical')

What of the social forces in support of the government? The events of February and October were disastrous for what was termed the 'mega-union'—an organization of union leaders accommodated to the neoliberal model, composed of militants of MNR, MIR, ADN, NFR, and UCS, the 'parties of the system.' While by no means defunct, this movement in its leadership is clearly in crisis, seriously questioned by the rank-and-file members in the base organizations. One aspect of this crisis is that in Mesa's distribution of positions the MNR, MIR, and NFR were entirely sidelined. As a result, for example, workers aligned with these parties have been 'ignominiously' expelled from a number of enterprises in which they had been installed. This has been the case, for example, in the National Highway Service (*Servicio Nacional de Caminos*) as well as COMIBOL, ECOBOL, COTEL, la Federación de Petroleros, and other public enterprises. However, it seems that the core leadership group (*el 'comando laboral'*) continues to function clandestinely under the '*batuta*' of David Olivares. In this context, the 'mega-union' still controls the *Confederación de Gremiales de Bolivia de Francisco Figueroa* (UCS), the *Confederación de Trabajadores Gráficos*, the *Confederación de Harineros*, *the Confederación de Trabajadores Municipales*, the MNR-controlled Confederation of Light and Force (COTEL), the *Confederación de Ferroviarios*, several mining unions linked to enterprises owned by De Lozada, several federations of *fabriles*, and several unions in the health sector.

The social forces so ranged behind and in support of the government cannot be viewed as particularly powerful, but nevertheless they constitute an important, if not critical, factor in the correlation of class forces. For one thing, the organizations led by Braaulio Rocha (MIR), Juan Melendres (MAS), and Mauricio Cori (MAS) have distanced themselves from the COIR de el Alto and COB. The problem is that these organizations are a critically important factor in El Alto and if they were not to join the social movement against Mesa and his neoliberal policies they would considerably weaken it. At the same time, rank-and-file members of these and other organizations in El Alto are putting pressure on their leaders with widespread questioning of their inclusion to 'negotiate' with the government. At this level, as they say, 'anything might happen.' One such example is what happened to Mauricio Cori, an organizer and MAS militant, on March 30, 2004. He was suspended by the neighborhood Associations that he represented because of 'union corruption' and his politics vis-à-vis the '*cuoteo de pegas*' (distribution of jobs) in the Prefecture.

The 'mega-union,' with its radical discourse, took advantage of this conjuncture by attacking some of the Bloc's leaders such as Miguel Zuvieta, executive secretary of the federation of Bolivian miners and thus a key supporter of Solares, himself a miner, repeating their demands that he be removed from his position; they were also very careful, in the same context, to refrain from challenging the government itself and this because of the not insignificant number of positions militants of their parties hold in so many public enterprises if not the government itself. The leaders of the mega-union, with one foot in the government and the other in the opposition, it is said, are good *'equilibristas,'* able to juggle contradictory positions on diverse issues. An example of this cited by several observers is Angel Villacorte, an NFR militant and leader of the Confederation of Bolivian Drivers. On the one hand, in various conjunctures he has opposed Mesa's rightist politics; on the other, he has sought agreements with leftist critics to put pressure on Mesa.

Mesa and the middle class

In parallel to the complex restructuring that has gone on in regard to the correlation of social forces, sectors of the middle class, composed of intellectuals, progressive impresarios, journalists, and the mass media have seen the need, an urgent political matter, to portray the Mesa government in different colors from those applied to De Lozada. Indeed, many have presented Mesa as 'progressive' and 'anti-neoliberal.' Others, with the clear intention to obfuscate the issues and disorient, confuse, and disarm the masses involved in the October uprising, have pointed to his considerable popular support and tried to convert him into the 'personality of the year.' It turns out that all of this activity on Mesa's behalf is not purely a matter of political principle. Like Lupe Cajías, the so-called 'anti-corruption tsar(ina),' a significant number of Mesa's friends as well as journalists and members of diverse social groups and NGOs now find themselves with important jobs in state enterprises such as Channel 7, Radio Illimani, the Bolivian Information Agency (ABI), and the new propaganda machine put together by Mesa. Thus it is that the mass media seem to have ended their campaign against Mesa's World Bank-friendly policies and voices of dissent have been stilled. No longer are the major critics of Mesa's policies on the public agenda of the major media. And the Church, the myriad of NGOs concerned with human rights, and the 'Defendant of the People'—all have come round to qualified, albeit mildly critical, support of the Mesa government.

MESA AND THE WORLD BANK

By some, if not all, accounts Mesa's macroeconomic policies are the same or worse than those of De Lozada in terms of neoliberalism.

To date, despite tacit support given him from progressive quarters, he has not moved an inch from his predecessor's policies and has been exceedingly submissive in the face of instructions from the IFIs, one of several reasons why the rank and file in many labor organizations are turning more and more against his government despite the accommodations of their leaders. The World Bank report *Country Assistance Strategy* is fairly clear on the line that Mesa is expected to toe. Some see it as a form of straitjacket that gives Mesa no room to maneuver, placing him in a similar position to that of Gutiérrez and Lula—or Kirchner for that matter. Others see the World Bank and the IMF as convenient constraints. After all Mesa, who has been described by the *Financial Times* as a 'man of the people' but by others as a *'cachorro de la bureacrcacia'* (runt of the bureaucracy), has filled his cabinet with believers in neoliberalism.

In any case, the report establishes that at the end of June 2004 and again at the end of December the Bank will evaluate whether or not and to what extent the government has implemented the mandated policies before providing the agreed $300 million credit. Article 73 establishes that 'Bolivia will receive an annual assistance of $150 million if and when there is compliance with the conditions agreed to by the government'. The report adds that after evaluating whether macroeconomic conditions are improving or worsening the bank will revise the key short-term macroeconomic objectives included as quantitative criteria to measure achievements in the government's economic program that has the support of the IMF. Further, to complement these short-term indicators the bank will also 'evaluate recent tendencies (over the past 24 months) and short, medium and long-term prospective (over the next three years) related to economic and financial indicators such as fiscal balance, short term and total debt . . .'

Most interesting perhaps is the apparent commitment of the government to the Bank for the obligatory 'sale of gas to the US'. This question of whether or not sell the country's gas reserves or to renationalize them is one of the most contentious issues, not only in the October uprising but of the continuing popular struggle. At the time of writing (May 26, 2004) the government is reeling from the pressures of a vast mobilization against the privatization and sale of gas and demands to nationalize it. The organizers of this mobilization declared the opening of a 'gas war' that pitted 'the people' against the government and its supporters both within and outside (the World Bank, the US, etc.).

DEMOCRACY WITHOUT SOCIAL MOVEMENTS:
THE POLITICS OF LOCAL DEVELOPMENT

The politics of adjustment and social change in Latin America has been polarized between the option of reform or revolution—or, with reference to the associated power struggle, between elections or the social movements. This has been the case for Bolivia as well. In the 1990s, however, Bolivia, by means above all of De Lozada's Law of Popular Participation, set in motion a series of concerted efforts to institute another modality of social change, a less overt form of politics: local development and municipalization (Ardaya, 1995; Booth, Clisby and Widmark, 1995; Rodríguez, 1995; Gannitsos, 1998; De la Fuente, 2001; Retolaza, 2003). Bolivia was by no means alone in this experience but it did constitute a laboratory of sorts in experimenting with diverse policies and institutional arrangements for ensuring the political conditions needed to implement unpopular policies and to cope with the inevitable side effects of the bitter medicine prescribed by the neoliberal model.

This new approach to the politics of adjustment (Grindle, 2000) was worked out in a series of high-level meetings between government officials, including De Lozada, at the time Estenssoro's Minister of Finance, and diverse representatives of the international community of financial institutions and development agencies, including the World Bank, the UNDP, and USAID. By all accounts (actually, we only have one, that of the Danish Development Agency) there were three major axes of strategic concern in these meetings—a concern that is reflected in the policy reform program constructed in the process: (1) *productivity competitiveness* (how to ensure the productive transformation of the economy); (2) *social integration equity* (how to ensure 'popular participation,' the missing link between efficiency and equity); and (3) *state action governability* (how to ensure the social and political sustainability of the required economic and political reforms).

As for the new policy agenda worked out in these meetings, it was in fact implemented, with diverse permutations, by different governments in the 1990s under the tutelage of the World Bank and the IDB, the two institutions that assumed primary responsibility in the region for ensuring the social and political sustainability of the neoliberal model.[22] The fundamental institutionality of the model was established by executive decree in 1985 but revised via a series of executive decrees in 1997 and 2000. During De Lozada's administration (1993–97) and his return (2002–3), the government's neoliberal agenda was advanced by administrative fiat, constituting a fundamental model for export.

The one major difference from policies instituted elsewhere, and the defining feature of the Bolivian model (the *Plan de Todos*), was the emphasis placed on 'popular participation' and the legal-administrative institutionality of local development based on the idea of 'productive municipality' (Ardaya, 1995).[23] The Administrative Decentralization Law, another lynchpin of this development model and a centerpiece of De Lozada's broader political project, was designed not only as a means of facilitating a local form of social and economic—and political—development but, together with its companion decree, the Law of Popular Participation, to ensure a more democratic form of governance, i.e. rule by social consensus rather than direct control by the central government. An intended effect of this reform was to focus the attention of the social movements from the national to local arenas of struggle.[24] Another was the institution of a good governance regime that would be sustainable both socially and politically (World Bank, 1994; UNDP, 1996; Blair, 1997; Zakaria, 1998).

The aim of the government's new policy agenda and political approach was twofold: to support the twin agendas of free markets and democracy;[25] and to draw attention (and oppositional political forces) away from national economic policy and the associated power structure and toward a community-based form of local development (and politics) orchestrated by local governments in strategic partnership with stakeholder associations and 'civil society organizations' (CSOs). In this strategic partnership, the developmental NGOs were assigned the role of mediating between the grass roots and the external donors, while private sector organizations, when and where they can be incorporated into the development process, are expected to provide capital and entrepreneurship. In this context, the promotion of local democratic institutions, or democratic governance, serves to establish a new regime of control that seeks the political stability necessary to attract various forms of capital even as it promotes municipalization, local autonomy, popular participation, and economic development (Miró, 1998; Kohl, 1999a, b).[26]

The most critical *political* role in this municipalization model of participatory development and this collaborative or partnership approach was assigned to the development NGOs. These NGOs constitute a relatively modest channel of ODA—on average around 10 per cent of total funds—but they are expected to make a significant contribution to what might very well be termed 'political development': (1) the self-capacitation of community-based organizations in regard to the process of economic and social development; and (2) the demobilization of class-based organizations and the social movement, turning them away from a politics of class confrontation toward a democratic politics of negotiation and dialogue.

In regard to this new form of politics ('no-power,' in the conception of John Holloway, Toni Negri, and other advocates of 'grass-roots postmodernism') the NGOs were first pressed into service in the 1960s, in the wake of the Cuban Revolution and widespread concern in governing circles that another Cuba could emerge. In this context NGOs were enlisted to provide front-line assistance in projects of 'integrated rural development.' In this connection the lay missionaries of the 'private voluntary associations' (PVOs), converted into front-line development agents, joined their religious counterparts ('Acción Católica') to help pacify the countryside and quench the incipient fires of revolutionary ferment—and, at the same time, to alleviate the conditions of abject poverty while promoting an alternative politics to that of the social movements.

In the 1960s the institutional and policy framework of this project was provided by the Alliance for Progress. But in the 1990s, under similar conditions but in a very different context, the NGOs were contracted to offer the rural poor and the indigenous communities of peasant farmers an alternative to the confrontational politics and mobilization tactics of the social movements. In exchange for direct action taken against the structure of political power by the social movements, as a means of thereby pressuring the government for a fundamental change in its economic policy, the NGOs were partnered with the external donors and local governments in providing the rural poor 'overseas development assistance' in the form of 'micro-projects'—or, 'micro-solutions' to the 'micro-problems' of the rural poor. As for their 'macro-problems' these are a matter of direct action by the social movements that have managed to escape the intermediations of the NGOs.

In the specific context of the anti-government mobilizations over the course of the MNR regimes established by De Lozada and Mesa the NGOs have continued to provide service to USAID and other agencies of 'international development assistance and cooperation.' For one thing, as pointed out by Ada Sotomayor, these NGOs have helped defuse various revolutionary situations by turning base communities away from traditional concerns with class issues such as land and labor, deflecting such concerns into concerns for issues of gender, alternative development, and local politics. In this connection, she notes that 'we are not gender-oriented. This is foreign. We deal with our own traditional practices (and concerns). All *gender related* projects have failed.' Another thing, Sotomayor adds, 'NGOs are funded to control funding. Money does not go down to the roots (but) it seeds corruption and divides groups within the community' (December 10, 2002).

The increasing responsibility for rural development taken by NGOs in the late 1980s and throughout the 1990s, within the parameters of the neoliberal state, has been the subject of considerable study— by Duran (1990) and Fisher (1993) among others. These and other scholars and analysts have pointed to the NGOs as key actors in the implementing of rural development projects within the framework of the Law of Popular Participation and diverse efforts to capacitate local governments to undertake local development (Molina, 1997).

There is little question in this process of helping the rural poor gain greater access to the natural, physical or financial resources which have always been viewed as critical factors of development. At issue, rather, is the provision of technical and professional services associated with the accumulation of social capital—the capacitation of local, community-based organizations to participate in local development projects and to absorb the poverty alleviation funds made available by the World Bank and other donor organizations. In the service of this project—to work (and seek improvements in the lives of the power) within the local spaces available within the power structure rather than challenging this structure—the NGOs have managed to co-opt a significant number of community (and even some national) leaders in the popular movement, turning them away from a politics of confrontation and mobilization toward a politics of negotiation, peaceful dialogue, and popular participation (good governance).

The Mesa regime has been particularly assiduous in building on De Lozada's strategy of alternative local development, municipal autonomy, and partnership with 'civil society.' The economics and politics of this approach to 'development' are reflected in an increased emphasis on municipalization.[27] First of all, the 1994 Law of Popular Participation (LPP) transferred 20 per cent of the national government's annual revenues to municipalities through a revenue-sharing program and created new institutions for local control and oversight of municipal spending. Secondly, the LPP called for participatory planning to determine the allocation of new revenues. By 1997 more than 15,000 grass-roots territorial organizations (OTBs) were formed and registered with the government (MDH-SNPP, 1997: 18). In this organizational context community members came together to define their needs and priorities, often with the critical participation of, and support from, the NGOs that were enlisted to this purpose of capacitation.

In many areas Municipal Development Plans (MDPs) were written by private contractors or by NGOs following government guidelines and with the mediated support of ODAs (Ayo et al., 1998; Kohl, 1999b, 2002). A 1998 study, published by the Vice Ministry of Popular Participation and Municipal Strengthening (VMPPFM), pointed to a number of critical problems in the implementation of popular participation. In at least one half of the municipalities in this study the grass-roots territorial organizations (the TBOs) did not participate in the formulation of the Annual Operating Plans (AOPs) carried out by the municipalities. In many cases the design and implementation of these plans were left entirely to NGOs (Kohl, 1999b). In other cases traditional party clientelistic politics prevailed, the community was politically divided, or local elites managed to co-opt the new institutional structures (Booth, Clisby and Widmark, 1996; Kohl, 1999a, b).

The municipalization of economic and social development is reflected not only in an enlarged role for the NGOs but in developments at the political level, particularly in the efforts of MAS to win representation and control over municipal councils and prefectures.[28] In 2002 oppositional forces representing indigenous communities and peasant producers made considerable gains in local politics and MAS in its current electoral strategy is focused on the municipal elections in October, intending to take over political control at this level. This *toma municipal* is the first step toward MAS's strategic goal of gaining state power. The 2007 presidential elections are the next step. The fact remains, however, that democratic participation and good governance, like local development, thus far have meant competition for limited resources—participation in decisions as to how to spend the meager poverty alleviation funds channeled into the rural communities through the NGOs at the local level. It has also meant increased internal divisions within the popular movement, a demobilization of its forces of resistance and insurgency, and a shifting of politics from the national to the local level.

A NEW CONJUNCTURE IN THE POPULAR MOVEMENT

The October uprising created the most significant political conjuncture in a struggle that can be traced back to the institution of neoliberalism in Bolivia in 1985. This conjuncture brought about a realignment of social forces and a change in the political dynamics of the struggle. Having installed himself in office, Mesa began by bringing out the carrot, co-opting a number of major social organization leaders. According to the La Paz Department of the Workers Central (COD) there is an easy explanation for the apparent social peace that settled in the May conjuncture (*EcoNoticias* /ARGENPRESS.info, April 5, 2004). It seems that the position of a number of leading community leaders changed in exchange for a 'quota of power' (a position within one of the state institutions). The case of the sudden 'suspension,' on March 31, of Mauricio Cori, Leader of the Federation of Neighborhood Associations of El Alto and a MAS militant, is an all too typical example of efforts by the government to co-opt movement leaders, dividing thereby the leadership from the rank and file.

In the same situation activists of the union movement engaged in the struggle against the government agenda vis-à-vis (i) the social pact; (ii) a referendum on oil and gas; and (iii) a proposal for a constitutional assembly point toward a u-turn in the position adopted by a number of organizations in October 2003 in the heat of the uprising. In this connection particular reference is made to sectors of the left, particularly MAS and including Morales himself, as well as journalists, the Catholic Church, representatives of diverse human rights organizations, and other middle-class groups that used the specter of an imminent military coup to justify their support of the government's plan and approach to dealing with the unresolved issues behind the February and October social explosions.

The position of Morales in this political context was particularly 'troubling' for the movement. Hitherto, as leader of the *cocaleros*, he had always been engaged in the struggle. But post-October he has taken a series of progressive steps backwards—to the right, to be more precise—in support of the government and its agenda and in direct opposition to, and in conflict with, the more confrontational stance taken by Quispe in El Alto and Solares through the COB. Not that there is anything surprising in this. It clearly reflects his commitment to electoral politics. It seems that Mesa and Morales needed—and still need—each other. Notwithstanding his government coalition, Mesa could not—and cannot –ignore or even govern without support from the second political force in the country (in the parliament to be precise). At the same time, Morales in his own estimation cannot achieve political power, at the municipal level in 2004 and the presidency in 2007, through mass mobilization. At the very least, if state power were to be achieved in this way it would be difficult, if not impossible, to sustain. And, it would seem, achieving state power is now Morales' first consideration, as it is for every political leader who has taken this road.

The harsh economic measures that the government is implementing, whether out of commitment or pressure from outside, constitute a sort of straitjacket if not for the government then certainly for the labor movement leaders who, for one reason or the other, have sided with the Mesa government. Demands from the rank and file for work, bread, a fair wage, education and health, and other basic needs are not being met by the regime. As a result, there is evidence of a slow and eventual political awakening among the rank and file. Although sporadic and isolated hunger strikes and demonstrations are back on the agenda of a protracted class struggle and a return of social conflicts that erupted in February and October 2003 but presently contained by various means. These conflicts will undoubtedly spill out of the constraints designed for it.

The government has sought to channel and at the same time contain the political forces in conflict via a plan to call for a referendum on the contentious issue of gas and set up a process leading toward the creation of a constituent assembly. At this point it appears that the political dynamics associated with these two issues, together with the municipal elections later in the year, will absorb the energies of the political class—and contain the political forces in conflict. But one never knows—certainly not political sociologists like ourselves. Popular movements seem to have their own dynamics which are difficult to predict or to understand precisely even after the fact. The events of February and October are cases in point. The 'developments' associated with them certainly exceeded the capacity of most observers to predict and for the leaders of the popular movement to control. All that these leaders could do under the circumstances was to 'jump on the bandwagon' of an ongoing struggle, or, to be more metaphorically precise, to jump into the wave of insurgent forces and mobilize these forces in some defined direction. As for the current conjuncture the issue on the immediate agenda is that of gas, an issue on which there is a clear division. On the government's side there is the proposal for a referendum, which has the support of, among others, Morales and the MAS. On the other side—against holding a referendum (because 'the people have already decided')—are the residents of Tarija where 80 per cent of Bolivia's gas reserves can be found, as well as the social movements associated with Quispe and Solares, principally in El Alto but also the broader labor movement led by COB.

In this post-October current situation, the positioning of diverse groups and organizations involved in the struggle for political power is critical. This is particularly the case for Morales vis-à-vis the MAS. Morales' own political behavior, and his critical support of Mesa's government of transition, characterized by an increasing distancing from the social movement, is supported by the following political analysis: if MAS were to promote a politics of social mobilization it would be playing into the hands of those sectors on the right seeking to destabilize the political situation so as to provoke a military coup. What has to be done is *not* mobilize the social movement but rather create a climate of peace and stability until the municipal elections in which MAS will undoubtedly win more than 100 municipalities and thus advance its political agenda; and, of course, the same applies to the longer-term objective of taking power at her national level in 2007.

This analysis—or rather, rationalization—of MAS's electoral strategy can be contrasted to an analysis that supports a contrary approach taken by, or in support of, the social movements.[29] In this analysis (Solón, 2004), 'For Bolivia to bet on the municipal elections is a serious mistake':

> The issues placed on the agenda today are not of a municipal character [and cannot be dealt with at the level of local politics]. The [critical] issues are national such as gas (preservation of reserves in the public interest), the viability of the neolibral model, the Constituent Assembly, impunity, FTAA, The Andean Free Trade Pact with the US . . . . To advance toward true (substantive) change requires the articulation of a broader and united movement mobilised on the basis of these [critical] issues. Only an articulation of this scale and scope that includes COB, MAS, CSUTCB, the neighbourhood associations and broad sectors of the [working] would be capable of stalling any attempt at a coup and at the same time moving away from the neoliberal model. (Solón, 2004)

In the context of this political analysis Mesa presented to Congress (on May 18) a proposal for a referendum on five highly contested political issues having to do with the ownership and disposal (public or private?) of the country's productive resources in oil and gas. The referendum, set for July 18 (later postponed to August), was announced in a context of continuing conflict, with teachers engaged in a series of violent confrontations with the government over salary issues as well as a demand to nationalize oil and gas, and COB (Solares), taking a similar position on the issue of oil and gas, disposed to 'negotiate [with the government] but in the context of continuing mobilizations' (*LJ*, May 20, 2004). Quispe, in the same context, announced his resignation from his parliamentary position to join the peasant movement in its opposition to the government's proposed referendum and its policies related to gas and oil. Within a few days (May 24) Quispe further announced his rejection of the government's popular referendum and the opening of a 'gas war' along the lines of the 'water war' of 2000. At the same time Roberto de la Cruz, leader of the Regional Workers Central in El Alto (COR), repeated the position that the referendum was not necessary because the people had already spoken in favor of the nationalization of gas. For his part, Rufo Calle, leader of the peasants of the La Paz Department, announced the end of a ten-day truce given the government on this issue and the beginning of a new round of mobilization in the form of highway blockades, a tactic adopted by a group of over 700 teachers and put into action in regard to highways connecting La Paz to Oruro and Lake Titicaca. The teachers announced that this action was but the first in a series of increasingly radical planned actions. At the same time the parents of their students in La Paz, Oruro, and Cochabamba went on a hunger strike in protest against the 15-day teachers' strike while the Minister of Oil and Gas Development announced his resignation in opposition to the referendum, while the government announced its intention to proceed as planned (*LJ*, May 26, 2004).

CONCLUSION

The popular movement in Bolivia has three main protagonists: organized labor in the major urban centers; the indigenous peasants and rural workers of the *altiplano* and the *cocaleros* of Chapare; and the community-based indigenous movement in its various divisions, most importantly in El Alto and less so in the eastern lowlands. Needless to say, the salient indigenous factor is not easily isolated from the issue of class central to both the peasantry and organized labor. The way in which these factors (ethnicity, peasant production, and labor) combine in the movement is difficult to determine precisely but responds to conditions that are in part internal and specific to each movement and in part shared in diverse historic conjunctures of what is predominantly a class struggle.

The political dynamics of this struggle are complex but as we see it they surround and arise out of the relationship of the movement to the state. The state is the dominant fact in the history of the popular movement and this primarily because it is the fundamental repository of political power, an instrument used by the dominant and ruling class to allocate the productive resources of Bolivia's society in its favor and interest. This point might be obvious, a conclusion all too easily drawn from an analysis that presupposes it. A confirmation of what is well understood, it might be said. But it bears restating and emphasis, given the diverse conceptions of power and the different paths to achieving it. Our conclusion is that electoral politics and the pursuit of political power via the instrument of a political party such as MAS is a dead end for the movement. Electoral politics is a game that the popular movement cannot win, governed as it is by rules designed by and that favor the dominant class, and that compel the movement to settle for very limited change and the illusion of power. Every single advance of the popular movement in Bolivia has been through a strategy of mass mobilizations.

Another pitfall in the way of social movements is local development and local politics. To all appearances local development provides the popular movement with a more viable path toward political power. For one thing power at this level is more democratic and easier to win or hold. For another it is closer to the community and can be achieved without confrontation with the power structure; indeed, it is predicated on a nonconfrontational approach, and offered to local communities as an option to the revolutionary path of social mobilizations. Local politics and alternative development provide at most a micro-solution to the community's micro-problems. Local autonomy and alternative development are achieved at the expense of substantive social change. It cannot and does not change the structure that allows the dominant groups to appropriate the lion's share of society's productive resources. This structure is shaped by national macroeconomic policies, the underlying power of which is in the central government. The politics of local development does not give the communities access to the natural resources, and the capital, both financial and physical, for them to bring about their own social and economic development.

In the diverse conjunctures of political developments that we have traced out it is possible to see the contours of a class and ethnic conflict that has its origins in a long history of colonial and postcolonial exploitation and oppression, and that in this context has the potential of mobilizing, if not uniting, a large part of diverse sectors of the popular movement. In no Latin American country are the conditions for a revolutionary situation so ripe, if not at hand—just below the surface we might say. At the time of writing (June 2004), the swell and rising tide of revolutionary ferment has receded but this could very well be the lull before the storm to come, which is to say, Bolivia presents us with the possibility of a truly revolutionary movement with all of its trials and tribulations, and messiness on the ground.

But to advance the popular movement in a revolutionary direction several conditions are required. One is that the popular movement needs to coalesce around a powerful organization of insurgent forces. The organization with the greatest potential in this regard, notwithstanding the failure of its most recent call for a general strike, or, for that matter, a failure to assume this role over the 50 or so years of its history, is the COB. A major source of this potential is COB's organization, a single nested structure of affiliation at the provincial, departmental (regional), and national level that has demonstrated a capacity for bringing together in collective action diverse sectors of the popular movement, in particular the country's peasantry and the working class.

Notwithstanding its historic failures, and a history of accommodations with existing governments and a propensity to internal crisis, the COB in the current conjuncture has the potential of constituting a critical mass of insurgent forces and mobilizing them into a movement that could potentially not only bring down the Mesa government but also change the course of Bolivia's history. However, it cannot play this revolutionary role alone; it must combine with other revolutionary forces, particularly those constituted by the *Aymara* indigenous proletariat of El Alto, organized by CTUCB and presently under the command of Quispe. A second condition is that the working class, led by COB, need to unite their struggle with the indigenous movement and the broader popular movement. MAS to some extent provides a political condition of such unity, COB less so in that it is precisely the division between different sectors of organized labor and the indigenous movement that has tended to and still divides COB, weakening its political responses to the government's macroeconomic policy. As for MAS, Morales, its leader, has pointed out in different situations that 'the workers [have always] believed themselves to be the "architects" of the revolution' and that the left in its representation of the class struggle 'talk of unity in diversity but have never put the experience of diversity into practice' (Morales, 2003). A third condition for a revolution is for the base organizations and insurgent forces in the popular movement to break away from the system (UCS, ADN, NFR, MIR, MNR) and its electoral politics. This is an escapable conclusion drawn from our analysis of decades of political dynamics in Bolivia.

The prospect for these developments is difficult to gauge. Some observers see this as 'not that difficult', while recognizing a stumbling block in the not insignificant number of union and movement leaders who are in dialogue with the government within the framework of a social pact or support the electoral path followed by MAS and the politics of a truce in the movement with the government. Others see an even greater obstacle in the diversion of the indigenous communities at the base of the popular movement into a politics of local 'autonomous' development and their adherence to an electoral strategy, not only in regard to the projected 'toma municipal' of MAS but the presidential elections in 2007. In this context, MAS is the repository of a large mass of forces of resistance but these are tied into the system via a politics of reform. Thus the social movement has to contend not only with the forces ranged behind the government but also the demobilizing approach of the most powerful political movement on the left.

In this situation, it would seem that, short of waiting for another revolutionary situation to arise, the best if not the only hope for the movement is for the rank and file to depose the leaders who are holding the movement back—to leave them behind. This might be possible in the case of the neoliberal parties of the system bound to a capitalist path to development and neoliberalism. But for MAS this is not so easy, or even possible, unless (as recently proposed by Escóbar in his battle with Morales) MAS is put under the control of a revitalized COB and thus subordinated to the broader popular movement as one of its major political instruments. But this scenario has various difficulties, not least of which is Morales himself. Having chosen the electoral path to power—the *toma municipal* in 2004 and the presidency in 2007—he has not only abandoned the dynamics of mobilization (the 'revolutionary path' to power, we might say) but any pretension of being a socialist let alone a revolutionary. He knows all too well that a commitment to play by the rules of electoral politics commits him to a capitalist path to national development should he, as he very well might, eventually be elevated to state power. This is one reason why he turned or swung to the right after October, concerned as he was to avoid any situation that might jeopardize the viability and survival of the institutions of the capitalist state. He fully intends to make good use of them in his bid for power and, it is clear, he is quite prepared to abide by its rules and limits.

As for Mesa it is abundantly clear that he is not the progressive reformer that some have made him out to be. Whether by ideological conviction, vulnerability to pressures from outside interests, ties to the dominant class, or simply political ambition, he and his regime can only be characterized as fundamentally neoliberal. Thus Mesa does not have the slightest intention of negotiating the end of a neoliberal approach to national policy, even if his government were able to ignore the inexorable pressures exerted by the World Bank, IMF and US—i.e. the imperial state. Mesa's predicament, as he himself prefers to see his situation, is how to balance these pressures and the requirements of good government (with sound policies, etc.) against the conflicting pressures of the popular movement. The irony is that without the nationalization of oil and gas reserves, a fundamental demand of the popular movement, and a policy of paying the external debt, a policy also opposed by the movement, and the restatification/ nationalization of the country's major public enterprises and capital funds (pensions, etc.), the government has no access to the financial resources needed to respond to these conflicting demands. And, like De Lozado, Mesa has sunk the country further into debt, the people into poverty, and the government into a relation of dependence on foreign capital.[30]

On this point, we conclude that the prospect for revolutionary transformation of the country's politics and economics is not far off. At the very least we can point to the likelihood that the contradictions that beset the ruling class and existing regime will continue to generate revolutionary situations of one sort or another. The question is whether or not the revolutionary left will be able to respond to the challenge. While this remains open-ended, several things are clear. One is the critical issue of state power—capture of the greatest repository of political power by the popular movement. Another is that the mass mobilization of insurgent forces, rather than electoral politics –the revolutionary road to state power—is the only viable method and means by which the popular movement can by its own actions bring about that 'new world' of social justice placed on the political horizon at the World Social Summit. In this regard Bolivia might yet surprise us.

6

Social Movements and the State: Political Power Dynamics in Latin America

Economic and social development requires changes in the structure of class relations and the configuration of political power. The question is how to bring these about. This question continues to bedevil social and political analysis, and politics, notwithstanding the plethora of sociological and political studies into, and decades of theorizing about, the political dynamics of class struggles and power relations in different contexts and conjunctures. Related questions that also remain unresolved include questions about the organizational form social change should take and the politics involved. Change at what pace and in what direction? On the basis of what agency and strategy? Despite the many theoretical probes into these dynamics such questions remain at issue in a series of theoretical and political debates on changes in class relations and political power. But what is clear is the dominant view that the road to social change and the transformation of class structures is paved with political power. Also, the central issue remains control of the state, the major repository of political power in regard to both the allocation of society's productive resources and the coercive power with which to enforce its decisions and policies.

With particular reference to developments in Latin America it is possible to identify three basic modalities of *social change* and *political power*. One is electoral politics—the pursuit of political power on the basis of political parties, which, as Max Weber noted years ago, were formed for this purpose. Another involves the construction of social movements. Unlike political parties social movements are not organized to pursue power as such. Although they are clearly engaged in the struggle over state power this struggle is an inescapable consequence of their quest for social change and anti-systemic politics of mass mobilization. A third way of 'doing politics' (social change via political power) in the context of developments in the 1980s and 1990s) entails social action in the direction of local development. This form of politics seeks to bring about social change (and a process of economic and social development, i.e. improvements in the lives of the poor) not through confrontation with the structure and agencies of political power but through the accumulation of social capital (the capacity of the poor to network and organize collectively) within the local spaces available within this structure. This concept of 'social capital' is central to and defines the dominant approach to social change in the mainstream of development theory and practice within the framework of the neoliberal model (Woolcock and Narayan 2000; Harris, 2001).[1]

While the electoral road to political power requires conformity to a game designed and played by members of 'the political class,' social movements generally take a confrontational approach to change and pursue a strategy of mass mobilization of the forces of resistance against the system and the political regime that supports it. In this context, structural and political dynamics are polarized between two fundamentally different approaches to social change and political power. The action dynamics of this political option—*reform or revolution* in the classical formulation, or *local development versus social movements*, in our recontextualized formulation—are not new. With diverse permutations they can be traced out across Latin America. What perhaps *is* new are the dynamics of change associated with the advance of an alternative 'social' or 'non-political' approach—a 'new way of doing politics' associated with the rise of grass-roots community forms of social organization and local development. The dynamics of this approach are a central issue in political developments across Latin America today. The authors have come to this conclusion on the basis of a systematic comparative analysis of the relationship between the state and social movements in four countries: Argentina, Brazil, Bolivia, and Ecuador. A summary of considerations that led us to this conclusion is presented in what follows.

THE CAPITALIST STATE IN LATIN AMERICA

The neoliberal model is predicated on a minimalist state—withdrawal of the state from the process of economic and social development and its replacement with the 'free market,' a structure supposedly freed from the constraints of government regulation and other interferences in the normal workings of a system in its allocation of society's productive resources (to determine 'who gets what,' or, in the language of economics, who secures an appropriate 'return' to each factor of production). Conditions for this retreat of the state emerged in the early 1980s, following the first round of neoliberal policies linked to the region-wide external debt and fiscal crisis. The first neoliberal experiments were led by military regimes in the southern cone of South America (Chile, Argentina, Uruguay) under conditions of a 'dirty war' against 'subversives' (unionists, political activists, etc.).

A second round of neoliberal reforms was implemented under conditions of 'redemocratization' (the rule of law, constitutionally elected civilian regimes, and the emergence and strengthening of 'civil society,' sphere of groups and associations between the family and the state). This round of 'reforms' allowed for and induced the widespread transfer of property, productive resources, and incomes from the working class and the mass of direct producers to an emerging capitalist class of investors and entrepreneurs. With the popular classes experiencing the brunt of the sweeping structural reforms associated with the 'new economic model' (neoliberal free market capitalist development) and bearing most of its social costs (social inequalities, unemployment, low income, social exclusion, and poverty), widespread discontent sparked several waves of protest movements directed against the system. The neoliberal policy regimes became ungovernable, generating pressures to move beyond the Washington Consensus. The outcome was the construction of a new policy regime—a neoliberal program of macroeconomic policies combined with a new anti-poverty social policy and the institutionality of a 'new economic model' (Bulmer-Thomas, 1996).

Parts of this model, such as a policy of administrative decentralization and a social policy that targets the poor for reduced public resources (a 'new social investment fund), were widely implemented in the 1990s. Other elements of this model, such as the municipalization of development and a system of 'democratic or local' governance based on the 'participation' of civil society ('stakeholders' in the development process), were experimented with on a relatively limited basis, primarily in Bolivia (Ardaya, 1995; Palma Carbajal, 1995; BID, 1996; Booth, 1996; Blair 1997).

These experiments constituted a third round of neoliberal policies but yielded few positive results in terms of economic growth and social development. By the end of the 1990s and into the new millennium economic growth rates across Latin America were far from 'robust,' a far cry from the prosperity and economic growth promised by the World Bank and the ideologies of neoliberal capitalist development. Indeed, ECLAC, a UN agency that over the years has led the search for an alternative to the (neo)liberal model, was compelled by the growing evidence of sluggish and negative growth rates and a propensity to economic crisis (in the late 1990s, after two decades of neoliberal reforms) to project 'a new decade lost to development.' Other erstwhile supporters of the new economic model were constrained to recognize the fundamental dysfunctionality of the neoliberal model and the need for fundamental reform of the reform process—to move beyond the Washington Consensus (Burki and Perry, 1998; Stiglitz, 2002).

As evidenced by the country case studies conducted by the authors, the neoliberal form of capitalist development is not only economically dysfunctional but profoundly exclusionary in social terms and politically unsustainable. A decade of state-led reforms has not fundamentally changed the Washington Consensus on macroeconomic policy. Nor has it changed the character of capitalist development in the region. Two decades of neoliberal reforms have resulted in deepening social inequalities, the spread of poverty and conditions of social crisis and disorganization. Even Carlos Slim, Mexico's major contribution to the Forbes billion dollar club, and one of the region's greatest beneficiaries of the neoliberal reform process, has joined the chorus of negative voices leveled against 'new economic model,' viewing it not only as dysfunctional in economic terms but also as inherently ungovernable. This conclusion, which, as it happens, a number of economists at the World Bank and other international organizations involved in the development project suspected all along, has fueled a widespread search for 'another form of development,' a decentralized and participatory form of local development based on more sustainable forms of 'democratic' or 'good' *governance* (World Bank, 1994; Dominguez and Lowenthal 1996; Blair 1997). The result has been a veritable flood of proposals and alternative models for bringing about 'development' on the basis of social capital, i.e. though the agency of 'self-help' of community-based or grass-roots organizations, with the assistance and support of partner institutions and 'international cooperation' for development (Woolcock and Narayan 2000).[2]

SOCIAL MOVEMENTS VERSUS THE STATE

A decade of efforts to give the neoliberal reform process in Latin America a human face has failed. But what is needed is not to a move 'beyond the Washington Consensus' toward face-saving reforms of the model. A redesign of the structural adjustment program is not the solution. What is needed is a social revolution that will change class relations, property relations, and the class character of the state. We establish this conclusion in the form of several propositions elaborated as follows.

1. Capitalism in its social and institutional forms is 'the enemy,' but in the current historical context the neoliberal state is the major locus of class struggle

State power is generally defined in terms of an 'authoritative allocation of society's productive resources.' But what sets the state apart from other institutions is control over coercive power, or, in the language of social science, its monopoly over the instruments of coercion and repression in its defined function of maintaining 'political order.' The state has a range of powers but, as demonstrated by our review of state-social movement dynamics in this volume, ultimately is backed up by force. This fact has been well established in theory and the social movements are all too aware of it in practice. In each case examined in this volume—Lula in Brazil, Kirchner in Argentina, De Lozada and Mesa in Bolivia, and Gutiérrez in Ecuador—the coercive apparatus of the state has been systematically directed against the social movements. In this context state coercion has not been a matter of last resort, as viewed by so many analysts in the liberal tradition—a justifiable exercise of state power. Coercion or repression is part of an arsenal of weapons used by the political class to control the movements—to weaken them in their struggle for change in policy or social transformation. It is the range of powers that defines the relationship of the state to the social movements.

This proposition is confirmed by a review of the dynamics that surround the relationship between the state and the social movements in Argentina, Brazil, Bolivia, and Ecuador. In the context of the political dynamics examined by the authors, the relationship of the state to social movements can be defined (and is structured) in terms of the following strategies:

(1)   setting up *parallel organizations* to class-based, anti-systemic organizations, such as peasant organizations and unions, that have non-confrontational politics;[3]

(2) *repression* of class-based organizations with an anti-systemic agenda under certain circumstances and where possible or necessary;[4]

(3) a process of dialogue and negotiating with representatives of class-based organizations with the capacity to mobilize forces of opposition and resistance (FARC in Colombia, MST in Brazil, EZLN in Mexico);

(4) *accommodating* the leadership to policies of economic, social, and political reform, often with the mediation of NGOs;[5]

(5) *pacifying* belligerent organizations on the basis of a reform agenda, a partnership approach, and a populist politics of appeasement and clientelism;

(6) *strengthening organizations within civil society* that have a reformist orientation and a democratic agenda, and weakening organizations with an anti-systemic agenda, a confrontational direct-action approach in their politics;[6] and, when all else fails:

(7) *incorporating* groups with an anti-systemic agenda into policy-making forums and institutions.

2. In the context of electoral politics mass parties are transformed into parties 'of the system'—pro-business, beholden to the 'Washington Consensus' on appropriate macroeconomic policy.

The best case of this proposition can be found in the transformation of the Workers Party (PT) from a 'party of the masses' into a 'party of big business.' This outcome is the result of the long-term, large-scale structural changes *within* the party and in its relationship to the state. The decisive shift in this case, as in the case of developments in Bolivia related to the *Movimiento Hacia Socialismo* (MAS), is from mass popular social struggles to electoral politics. In this evolution the PT became an 'institutional party,' embedded in all levels of the capitalist state, and attracting as a result a large number of petit-bourgeois professionals (lawyers, professors, journalists), trade union bureaucrats, upwardly-mobile ex-guerrilla, ex-revolutionaries recycled into the electoral arena. A process of substitutionism takes place. Here the electoral apparatus replaces the popular assemblies, elected officials displace the leaders of the social movements, and the institutional maneuvers of the national political leaders in congress substitute for the direct action of the trade union and social movements.

The historical and empirical data demonstrate that elitist electoral leaders *embedded* in the institutional structures of the capitalist state end up competing with the other bourgeois parties over who can best administer the interest of the foreign and domestic, agrarian and financial elites. But the fundamental change in the shift of mass parties toward electoral and institutional politics is found in its class composition: it tends to become the party of ambitious *upwardly-mobile* lower-middle-class professionals whose *social reference* is the capital class. Both political developments in Brazil and Bolivia provide evidence for this assertion. Behind these developments can be found a change in *class consciousness* which reflects a change in *material conditions* of the elected politicians. Under these conditions, depending on the capacity of these leftist labor-oriented politicians to garner voting support among the working class, landless workers and urban *favelados* become bargaining chips to negotiate favors with big business.

The 'new class' of electoral politicians tends to look *upward* and to their *future* ruling-class colleagues, not *downward* and to their former working-class comrades. A similar development seems to occur within the trade union movement in its relationship to the state. In the case of Brazil under Lula, for example, upwardly-mobile trade union officials look upward to becoming congressional candidates and ministers, or administrating pension funds rather than downward to organizing the unemployed and the urban poor in general strikes with the employed workers. The PT's transformation into a party of international capital was accompanied by the transformation of the major trade union confederation (the CUT) from an independent, class-based union into more or less an appendage of the Ministry of Labor. The CUT in this context followed the PT along the path of 'state institutionalization' and 'substitutionism' as the national leaders pre-empted the factory assemblies in making decisions and relocate activities from the streets to the offices of the Ministry of Labor. Thus the parallel transformation of the PT and CUT avoided any rupture between them, a development that has not been followed in Bolivia. However, there are political forces in Bolivia that follow the Brazilian example.

The key theoretical point from this analysis is that the bourgeoisification of 'working class' or 'socialist' parties is not the inevitable consequence of globalization. Rather, it is the result of changing class ideology, the internal dynamics of party politics— changes that lead to institutional assimilation and, ultimately, subordination to the dominant sectors of the ruling class. This conclusion points to the profound limitation of electoral institutional politics as a vehicle for social transformation or even consequential reforms. Social transformation is far more likely to occur from the direct action of independent class based social political movements oriented toward transforming the institutional basis of bourgeois state power.

3. Electoral politics is a trap designed to demobilise the forces of resistance and opposition

This proposition is amply demonstrated by political developments in every country case study we have examined, and particularly in Bolivia, Brazil, and Ecuador. In their efforts to advance the struggle for political power many social movements seek a strategic or tactical alliance with electoral political parties, as with the MST and the PT, or, as with CONAIE in the case of Ecuador (Pachakutik) or MAS and MIP in the case of Bolivia. In this alliance social movements evolve and are transformed into a political instrument for the purpose of influencing regime policy within the system. As regards the outcome of this evolution the conclusion is clear. It is invariably at the expense of the popular movement, whose forces of resistance and opposition rather than being brought to bear against the power structure are dissipated and demobilized. Pachakutik in Ecuador provides a good example of this development, but to appreciate its theoretical and political significance we cannot do better than turn to Bolivia.

Our studies into ostensibly 'progressive' regimes with links to the social movements and neoliberalism suggest that electoral regimes, no matter what their social base or ideological orientation, inevitably became integrated into, and subordinate to, the imperial system. The result is that social movements and their members are prevented from achieving even their minimum goals.

Take the case of Ecuador. The petroleum workers and CONAIE, through its electoral arm Pachakutik, in the conjuncture of political developments that followed the 2000 indigenous uprising, entered into an electoral alliance with Lucio Gutiérrez and his *Sociedad Patriotica*. Upon taking office, catapulted into power on the basis of this alliance, Gutiérrez embraced a policy of privatization of petroleum as well as the policies of the IMF, ALCA, and *Plan Colombia*. Gutiérrez repressed the petroleum workers and turned the government's back on the indigenous movement (betrayed it, in the conception of CONAIE at its national congress in 2004). The result was a very weakened Petroleum Union, a discredited Pachakutik, and a seriously weakened and divided CONAIE.

In Bolivia, Evo Morales, the leader of the *cocaleros* and the Movement to Socialism (MAS), like so many electoral politicians and parties, after some advances, turned to the right. In the wake of the October 2003 uprising, in which he was notably absent, Morales supported the pro-imperialist, neoliberal regime of Carlos Mesa, playing a major role in dividing and attacking any large-scale mobilizations in favor of nationalizing petroleum. Our analysis of diverse electoral regimes suggests that this development is not in the least surprising. It is built into electoral politics, the inevitable result of its dynamics. In the case of Morales, his politics is undoubtedly geared to his quest to win the 2007 presidential election, a prospect that many analysts today see as increasingly unlikely given the mechanics of Bolivia's electoral politics (the need for a second round of voting should no candidate achieve over 50 per cent in the first round).

Political developments in Brazil suggests that no matter what the situation or electoral prospects, once a popular movement turns to electoral politics it is constrained to play by the political rules that sustain the dominant model and, in the current context, the neoliberal agenda. The MST in this context has not suffered the same debacle as CONAIE, because only a few members were in the government and it managed to retain a sufficient degree of autonomy to sustain the loyalty of its members. Also, unlike the *cocaleros* in Bolivia, the MST is not dominated by a single, electorally ambitious personality and is sufficiently grounded in class politics to avoid becoming a tool of the bourgeois state. Nevertheless, the MST's confidence in Lula and ties with the 'left' of the PT has undermined its opposition to Lula's reactionary attack on pensions, minimum wage, the IMF pact, and military support of US colonial occupation of Haiti. The danger here is that by continuing to give 'critical' support to a discredited regime, the MST will suffer the same discredit, a lesson that CONAIE has learnt all too well.

4. Local development provides micro-solutions to micro-problems, designed as a means of eluding a confrontation with the power structure and substantive social change

The best examples of this proposition are found in Bolivia and Ecuador. This is in part because of the 'indigenous factor' in the national politics of these two countries. In both cases indigenous communities have demonstrated the greatest capacity for mobilizing the forces of resistance and opposition, organizing some of the most dynamic social movements in the region. For this reason the World

Bank and the IDB targeted the indigenous communities and the social movements based on them as the object of what amounts to an anti-insurgency strategy: local development in the form of micro-projects of poverty alleviation.

Given the established dysfunctionality of the neoliberal model, and the tendency of this model to undermine democracy and generate destabilizing forces of resistance in the form of social movements, the architects and guardians of the 'New World Order' have turned towards 'local development' (micro-projects) as *the* solution (to the neoconservative problem of ungovernability). The World Bank in this context finances nongovernmental organizations (NGOs) within the 'third sector' of 'civil society as agents of 'local development' and 'good governance'—to combat the emergence of mass movements.

The first step in this strategy was to establish at the level of the state an appropriate institutional-administrative-legal framework. The next step was to enlist the services of the NGOs, converting them into front-line agents of the 'development project' (poverty alleviation) and, in the process, into missionaries of micro-reform. The NGOs provided the imperialist organizations cooperating in the development project entry into the local communities. The micro-reforms and NGOs promoted a pacific or 'civil' (non-confrontational) form of politics, turning the rural poor away from the social movements into local self-help 'projects' funded (and designed) from above and the outside. It also created local conditions for an adjustment to the discipline of globalization and its governance requirements—to create local conditions of imperial governance. In this context Heloise Weber (2002: 146) could write of micro-finance and micro-credit as a 'coherent set of tools that may facilitate as well as govern the globalization agenda.' From 'the perspective of the architects of global development,' she adds, 'the micro-credit agenda (and thus, the "poverty alleviation" strategy of the World Bank—"Sustainable Banking with the Poor") . . . is conducive to facilitating policy changes at the local level according to the logic of globalization . . . while at the same time advancing its potential to discipline locally in the global governance agenda' (Weber, 2002: 146). With reference to these developments we can well conclude that the official discourse on 'civil society' is little more than an ideological mask for an imperialist agenda—to secure the political conditions for neoliberal capitalist development.

5. Mass mobilization is the revolutionary way to political power and the only route to social change

Political developments in every country case study examined by the authors confirm what has long been a truism in Marxist class analysis. At issue in the class struggle is political power in the form of the state. Each advance in this struggle has been associated with mobilizational politics, while the recourse to electoral politics in each case has perpetuated the status quo.

In no Latin American country are the conditions for revolution as well developed as in Bolivia. At the time of writing (June 2004), the swell and rising tide of revolutionary ferment has receded but this could very well be the lull before the storm to come; which is to say, Bolivia presents us with the possibility of a truly revolutionary movement with all of its trials and tribulations.

However, to advance the popular movement in a revolutionary direction certain conditions are required. First, the popular movement needs to coalesce around a powerful organization of insurgent forces. In the case of Bolivia the organization with the greatest potential in this regard is the COB, which has uniquely managed to represent politically and advance the interests of both organized workers and indigenous peasants. A major source of COB's political potential is its organizational structure, a single structure of affiliation at the provincial, departmental (regional), and national level with a demonstrated capacity for bringing together and concentrating the collective action and mobilizations of diverse sectors of the popular movement.

Notwithstanding its historic failures and limitations, the COB in the current conjuncture has the potential of constituting a critical mass of insurgent forces and mobilizing them into a movement that could potentially not only bring down the Mesa government but also change the course of Bolivia's history. At issue here are three factors. One is the form of organization; another is leadership. The third critical factor is an appropriate and effective strategy and associated tactics, particularly as regards selected forms of struggle and 'the strategic alliances with popular, social, progressive and patriotic sectors that Luis Macas, the current leader of CONAIE, observes 'are necessary and fundamental for the process of revolutionary change' (La Jornada, Ojarasca, March 2005: 3). In this regard, the COB has to combine with other revolutionary forces, particularly those constituted by the Aymara indigenous proletariat of El Alto, and presently under the command of Felipe Quispe. Another requirement is that the working class, led by COB, need to unite their struggle with the indigenous movement and the broader popular movement. To some extent MAS provides the political conditions for such unity; COB less so in that it is precisely the division between different sectors of organized labor and the indigenous movement that still divides COB, weakening its political responses to the government's macroeconomic policy.

In mid-March 2005, with the Mesa government besieged by the popular movement against the privatization and denationalization of the country's strategic resources, some conditions for this political unity were finally available. For the first time in the post-2003 class and popular struggle, despite the serious divisions of interest and politics (of ethnicity and class), the *Aymaran* social forces under the direction of and mobilized by Quispe and those aligned with Evo Morales and MAS, the COB under Solares, the *junta vecional de Alto* and the impoverished middle classes have come together politically in opposition to the Mesa regime and its coalition of *empresarios* and center-right parties with US backing. In this context (in support of a new law promising a royalty rate up to 50 per cent and a corporate tax rate of 32 per cent) the social movements in Bolivia are conducting politics and making public policy on the streets, the *barrios* and in the countryside. In the face of the evident collapse of the political apparatus and state power, diverse groups and organizations in the popular sector of Bolivia's civil society are in the process of constructing a popular form of power, which, as Luis Macas with regard to Ecuador has pointed out, 'is in the commune [and found 'below, not above']. There is no questing here or there of an amorphous 'multitude' *à la* Negri or of the rejection of power *à la* Holloway. What we see are the political dynamics of popular power formation.

The prospect for these developments is difficult to gauge. Some observers see this as 'not that difficult.' Nevertheless they recognize the formidable obstacles in the significant number of union and movement leaders in dialogue with the government within the framework of a social pact. Other union leaders support the electoral path followed by MAS and the politics of a truce between movement and the government. Others see an even greater obstacle in the diversion of the indigenous communities at the base of the popular movement into a politics of local 'autonomous' development and their adherence to an electoral strategy, not only in regard to the attempted takeover of municipal government by MAS but the

2007 presidential elections. In this situation, MAS is an important repository of oppositional forces, but these forces are tied into the system via an electoral politics of incremental reform. Thus the social movement has to contend not only with the forces ranged behind the government but also the demobilizing approach of a powerful political movement on the left.

Perhaps the best if not only hope for the movement is for the rank and file to depose the leaders who are holding the movement back—to leave them behind as it were. This might be possible in the case of the neoliberal parties of the system bound to a capitalist path to development and neoliberalism. But for MAS this is not easy, or even possible, unless as proposed by the former MAS Senator Filomen Escóbar, in his battles with Morales, MAS is placed under the control of a revitalized COB and thus subordinated to the broader popular movement as one of its major political instruments. But this scenario has various difficulties, not least of which is Morales himself. Having chosen the electoral path towards power—the *toma municipal* in 2004 and the presidency in 2007—he has seemingly not only abandoned the dynamics of mobilization (the 'revolutionary path' to power, we might say) but any pretension of being a socialist, let alone a revolutionary. He knows all too well that a commitment to play by the rules of electoral politics commits him to pursue a capitalist path to national development should he, as he very well might, be elevated to state power. This is one reason why he turned or swung to the right—to avoid any situation that might jeopardize the viability and survival of the institutions of the capitalist state.

As for Mesa it is abundantly clear that he is not the progressive reformer that some have made him out to be. Whether by ideological conviction, vulnerability to pressures from outside interests, ties to the dominant class, or simply political ambition, he and his regime have to be characterized as fundamentally neoliberal. Thus he has no intention of negotiating the end of a neoliberal approach to national policy, even if his government were able to ignore the inexorable pressures exerted by the World Bank, IMF and US—i.e. the imperial state. Mesa's predicament, as he himself sees it, is how to balance these pressures, and the requirements of good government (with sound policies, etc.), against the conflicting pressures of the popular movement.

On this point, we conclude that the prospect for revolutionary transformation of the country's politics and economics is not far off. We can point to the likelihood that the contradictions that beset the ruling class and existing regime will continue to generate revolutionary situations of one sort or another. The question is whether or not the revolutionary left will be able to respond to the challenge provided by these situations. While this remains open-ended, several things are clear. One is the critical issue of state power—capture of the greatest repository of political power. Another is that mass mobilization of insurgent forces, rather than electoral politics—the revolutionary road to state power—is the only viable method and means by which the popular movement can by its own actions bring about substantive social change.

6. Social movements have failed to respond to the revolutionary challenge

Marx a long time ago argued that the capitalism in its advance creates its own gravediggers—a working class aware of its exploitation, disposed to overthrow the system. However, he noted, this development requires a revolutionary situation, the conditions of which are objectively given and subjective—structural and political. In several countries examined in this volume, particularly in Bolivia and Ecuador but also Argentina, these conditions have come together a number of times in diverse conjunctures: December 19/20, 2001 in Argentina, October 8–19, 2003 in Bolivia, and January 2000 in Ecuador.

To date in Brazil no such conjuncture has materialized. But at the same time, the presidential elections that brought Lula to power did create the opportunity for a new regime to use this power to bring about a social transformation. But this would require a socialist regime and Lula's regime is anything but that. In fact, we conclude that a socialist regime cannot take state power like this. Electoral politics binds any party to the system, turning it toward neoliberalism— toward forces that govern the system. Thus, as in the other cases examined in this volume, the 'moment' of state power as it were—and the 'opportunity of mobilizing the forces of resistance against the system—was lost. In the case of Brazil, the reasons for this were predictable given Lula's politics and the class nature of his regime.

In Argentina the struggle for political power has taken a different form. What emerges from the extended and massive popular rebellion is that spontaneous uprisings are no substitute for an organized political movement. The social solidarity formed in the heat of the struggle was impressive but short-lived. Little in the way of class solidarity reached beyond the *barrio*. The parties on the left and local leaders did little to encourage mass class action beyond the limited boundaries of geography and their own organization. Even within the organizations, the ideological leaders rose to the top, not as expressions of a class-conscious organized base but because of their negotiating capacity in securing work plans or skill in organizing. The sudden shifts in loyalties of many of the unemployed—not to speak of the impoverished lower middle class—reflect the limitations of class politics in Argentina. The *piquetero* leaders rode the wave of mass discontent and lived with the illusions of St Petersburg, October 1917, failing to recognize that there were no worker soviets with class-conscious workers. The crowds came and many left when minimum concessions were made in the form of work plans, small increases and promises of more and better jobs.

As in the other contexts studied in this volume the domestication of the unemployed workers' movement is located in a number of regime strategies. Kirchner in this connection engaged in numerous face-to-face discussions with popular leaders, making sure that the best work plans would go to those who collaborated with the government while making minimal offers to those who remained intransigent. In this context he struck an independent posture in relation to the most outrageous IMF demands while making concessions on key reactionary structural changes imposed by his predecessors. *Lacking an overall strategy and conception of an alternative socialist society*, the majority of the *piquetero* movement was manipulated into accepting micro-economic changes to mitigate the worst effects of poverty and unemployment, without changing the structure of ownership, income, and economic power of bankers, agro-exporters, or energy monopolies. The resulting political situation, played out with diverse permutations across Latin America, was a variation on the all too dominant theme of local development and reform—and a politics of negotiation and conciliation.

The problem with this style of politics is that the question of *state power* is avoided. In the specific context of Argentina it was simply a declaratory text raised by sectarian leftist groups who proceeded to undermine the organizational context in which the challenge for state power would be meaningful. In this they were aided and abetted by a small but vocal sect of ideologues who made a virtue of the political limitations of some of the unemployed by preaching a doctrine of 'anti-power' or 'no-power'—an obtuse mélange of misunderstandings of politics, economics, and social power. The emergent leaders of the *piquetero* movement, engaged in valiant efforts in raising mass awareness of the virtue of extra-parliamentary action, of the vices of the political class, were unable to create an alternate base of institutional power for unifying local movements into a force that could confront state power.

What is clearly lacking in this and other situations is a unified *political organization* (party, movement, or combination of the two) with roots in the popular neighborhoods, capable of creating representative organs that promote *class consciousness* and point toward taking state power. As massive and sustained as the initial rebellious period (December 2001–July 2002) was, no effective mass political party or movement emerged. Instead, a multiplicity of localized groups with different agendas soon fell to quarrelling over an elusive 'hegemony'—driving millions of possible supporters toward local, face-to-face groups that lacked any political perspective. Under these circumstances the forces of opposition and resistance were dissipated and the wave of revolutionary ferment receded.

Viewing these issues retrospectively leads us to the conclusion that is entirely consistent with the evaluation made by many activists within the movement: that it is a political mistake to seek state power from within the system—to turn toward electoral constitutional politics and join the government. This much is obvious. Assessments of the state–movement dynamic in other contexts have produced the same conclusion. The problem is that this conclusion does not get us too far.

Mobilizing the forces of opposition and resistance against the system is part of the solution—in fact, a large part, given the limits and pitfalls of electoral politics. This is clear enough. Indigenous leaders, like Humberto Cholanga of Ecuador, on the basis of struggles within the social movement, have embraced a class perspective on the 'indigenous question.' But another part of the solution is to create conditions that will facilitate the birth of a new revolutionary political party oriented toward state power. We can be certain that the process will be fraught with difficulties and will require the leadership of conscious political cadres. A close look at the experiences of the four countries provides answers to the limitations of social movements and electoral politics.

7. Socialism remains on the horizon—of the social movements

The socialization of the means of production and an egalitarian distribution of goods and services have been implicit and explicit goals of the mass social movements. The ritualistic declarations by the leaders of these movements of the vague slogan 'another world is possible' has failed to define a political direction and economic strategy that links popular needs with fundamental economic structural change. Faced with the growth of large-scale agro-export enterprises, 'agrarian reform' can be consummated only through social collective ownership of the means of production, as the MST has recently acknowledged and collective ownership and control of the country's strategic resources, as emphasized by the indigenous movements in Bolivia and Ecuador in the course of their protracted and ongoing struggles.

The return to power of the financial elite in Argentina demonstrates that state 'regulation' is incapable of directing capital toward large-scale, long-term investment in employment-creating economic activity. Only on the basis of a publicly owned financial and banking system and the domestic market, is it possible to design and implement national policies which could sustain a form of economic and social development in which the society as a whole, not just the few can benefit. Our study demonstrates that the 'national' bourgeoisie, including those oriented to the local market, are addicted to the accumulation of their capital, directing their profits to overseas accounts, recycling their earnings into the financial sector, and intensifying exploitation of their workforce, rather than expanding employment and reinvesting productivity in the home market. The necessary conclusion is that sustained and comprehensive industrial growth on a national scale requires public ownership under the control of employed and underemployed workers and professionals. That is, it requires a reversal of the privatization policy that has dominated politics in the region for well over a decade. The crisis of electoral elite politics—riddled with corruption, beholden to foreign creditors, immersed in the politics of privatizations—can be resolved only by a transition to democratic collectivism, which prioritizes political control from below, productive investment over debt payments, and a recovery of the strategic sectors of the economy.

There is a growing popular dissatisfaction with the endless social forums, vacuous declarations and ritualistic self-congratulations that have become a substitute on the Left for organizing mass struggle based on a clear albeit implicit socialist program.

Throughout our five years of fieldwork in the four countries with hundreds of unemployed and employed workers, in the formal and informal labor market, among downwardly-mobile public employees and underemployed professionals, among Indian leaders and activists, we have found a much clearer option for a socialist transformation and rupture with the electoral political class than among the professional Social Forum attendees who still live in the Tower of Babel of diversity and dispersion of ambiguous formulas, rather than concrete actions in the direction of socialism.

To be sure, the socialism of the social movements is not the state socialism of the former soviet socialist republics of Europe. It is a socialism that is grounded, as Macas, the current leader of CONAIE, has it, in 'our communities existing in collectivity', that is, 'in a communal world' that reflects '["places at the center"] principles [and relations] of redistribution ... reciprocity [and] complementarily in the face of the competition that is killing us' (*La Jornada*, Ojarasca, March 2005: 3). 'We live daily a structure, a power, an excluding and hegemonic institutionality [neoliberal capitalism]', a 'rotten [oligarchic] power' which needs to be transformed 'from below' into a power that 'we are constructing in ... unity'. The institutional form of this power, he adds, will be 'a new pluri-national state'.

CONCLUSION

Historians of development theory have written of a counterrevolution in development theory and practice traced back to the exhaustion of the Keynesian model of state-led economic development and the appearance of a 'new economic model' based on the neoclassical doctrine of free world market as a fundamental engine of economic growth as well as the most efficient mechanism for allocating productive resources across the system, essentially replacing governments in this role. Other historians have identified a paradigmatic shift traced back to a 'theoretical impasse' brought about by a structuralist approach to social analysis and a political project to create a form of society characterized by a fundamental equality in social relations and equity in access to, and distribution of, the world's wealth. As in Marx's day these intellectual developments 'appeared' and were analyzed by many as a war of ideas, a struggle by different ideas to realize themselves. However, as Marx understood so well in a different context, this conflict in the world of ideas reflected conditions of a class struggle in the real world, namely actions of working peoples across the world to improve their lot through a process of social change. The 1970s saw a new conjuncture in this struggle: a counteroffensive launched by capitalists and their ideologues and state representatives against the working classes, seeking to arrest and reverse the gains achieved over two decades of economic, social, and political development.

The new economic model of neoliberal policy reforms was one major intellectual and ideological response, a major weapon in the class war unleashed against the popular classes. Another such response took the form of a sustained effort to disarm the poplar movement, to disarticulate its organizational structure and class politics, to turn the popular movement away from its struggle for social change and state power for control over the major repository of political power. The aim here was to construct a new modality for achieving social change based on a new way of doing politics, namely to take the path of 'anti-' or 'non-power'; to rely on social rather than political action in bringing about social change without a confrontation of the power structure; to seek change and improvements within the local spaces available within this structure; to from partnerships with other agencies in the project of local development to empower the poor to act in their own lives, participate actively on their own development and the good governance agenda, without challenging the larger structure of economic and political power.

The conclusion that we draw from an analysis of this modality of social change and the political dynamics of the development project in Latin America is inescapable. The only way forward for the working classes and the popular movement is political power: to abandon the development project and engage the class struggle—to directly confront the holders of this power and contest power in every arena open to it. However, as in earlier political conjunctures, there are two roads to state power, both fraught with pitfalls: the road of electoral politics and a revolutionary politics of mass mobilization. Perhaps Morales, the leader of MAS and erstwhile leader of the *cocaleros*, the coca-producing indigenous peasants of Chapare, best exemplifies the dilemma (and the difficulties in pursuing both paths at the same time). Morales' decision to take the electoral road to state power (to gamble on his chances of winning the 2007 presidential elections) was largely responsible for defusing the revolutionary situation created by insurrectionary politics of the popular movement in October 2003. Mesa would never have come to power without the support of Morales. In that context (he did not play a major role in this insurrectionary politics, in the bloody street protests that forced Gonzalo Sánchez to resign and seek exile in the US), the weak Mesa government would never have survived without support for Mesa's call for a referendum on the explosive gas issue or his continuing support. As Eduardo Gamarra (*The Herald*, January 15, 2005: 5A) noted: 'The length of Mesa's tenure [in office and power] is largely due to Evo's supportive role.' At the same time Morales himself continues to be pressured from the left of the popular movement, compelled to respond to its more radical politics. For example, Morales continues to insist that he wants to win power only through the ballot box[7] but in January 2005 was constrained by the politics of the radical left to demand that Mesa resign and call an early election unless he rolls back recent gas price increases. At issue for Morales was how to maintain his position in the popular movement while finding himself standing on the sidelines of the class struggle during anti-government strikes that shut down the cities of El Alto and Santa Cruz. The radical left in this context is rejecting elections as a means of achieving state power, but Morales is holding to the tough line of electoral politics while at the same time being forced to respond to the radical politics of the revolutionary left. Alvaro García, a university professor who is a part of this left but at times serves as an advisor to Morales (he has severed ties to most of the radical left), in this political context observes that '[w]hen the radicals are powerful he moves towards them.' The point is, he adds, Evo 'fears that he will lose his base of support to the more radical elements.' And well he might. It was only after a meeting (in mid-March) with the six *cocalero* federations that he (rather than *cocalero* leadership) decided to 'radicalize their actions' (*La Razon*, March 20, 2005)

The response of Morales to radical politics tends to be tactical rather than strategic. It ties into a general conclusion that we have drawn from our analysis of diverse class struggles and the politics of social change. The inescapable conclusion is that a radical politics of mass mobilization is an indispensable condition for advancing the struggle for social change—to bring about a new world of social justice and real development based on popular power (control of working peoples of the state). In practice it is necessary to combine both electoral and mass revolutionary politics—as Morales is discovering. Just a week before this writing (March 20, 2005), Morales had a series of meetings with the *cocaleros* about forming the social base of the movement that he still leads as leader of MAS. Whatever transpired at these meetings it was clear that Morales returned to Congress with a renewed sense and commitment to a politics of social mobilization, threatening to radicalize the politics of popular resistance if the Senate chose not to pass the new congressional oil and gas (hydrocarbon) law requiring oil companies exploiting the country's strategic resources in the sector to increase royalty payments from a low of 18 per cent up to 50 per cent. 'We [the *cocaleros*] have changed our methods of struggle,' Morales declared, 'from blockades (as a tactic of pressure on parliamentary action) to effective and emblematic means required by patriotic defense' (*La Jornada*, March 21, 2005: 30). But a mobilized people is the *sine qua non* of revolutionary change—and revolutionary change is the only solution.

Notes

PREFACE

1. *Observer Worldview* (Ed Vulliamy, April 21, 2002) reported that plotters, led by Venezuelan businessman Pedro Carmona, who tried to topple President Chavez from power in mid-April 2002, were closely linked to, and encouraged in their plot by, senior officials in the Bush administration, particularly Elliot Abrams, complicit in the infamous Iran/Contra Affair under President Reagan, and Otto Reich, one of Bush's chief policy-makers for Latin America.

CHAPTER I

1. One of the first to recognize and define this problem was Samuel Huntington, who, in 1975, together with two trilateralist colleagues, submitted a report to the Trilateral Commission (Crozier et al., 1975) which viewed democracy as a seriously flawed system given its tendency to generate expectations and forces of radical change that cannot be contained within the system. Ten years later Robert Kapstein, Director of the US Council of Foreign Relations, one of Washington's critically important foreign policy forums concerned with the project of constructing a New World Order (or in the language of neoconservatism, 'the new imperialism'), raised the specter of political instability and ungovernability in the context of a trend toward excessive social inequalities, growing poverty, and the extreme polarization of world society (Kapstein, 1996). The World Bank in particular took this question seriously, seeing it as central to the development enterprise and its mandate—to alleviate poverty, via a less exclusionary process and if necessary by means of improving access of the poor to society's productive resource (World Bank, 1994). The following year, however, ten years upon Kapstein's published article in *Foreign Affairs*, saw the emergence of another problem that assumed crisis proportions in Asia in mid-1997: the Asian Financial Crisis. The problems associated with this crisis put the question of governance on the international agenda as a matter of urgency: to re-regulate or control the vast pools of volatile capital, predominantly in the form of portfolio investments, by means of a 'new financial architecture' (on this see, *inter alia*, Stiglitz, 2002). More recently, the issue of governance has been re-examined in the context of a revived concern that the polarization between the poor and the rich in world society and the global economy is threatening to undermine democracy and create political instability (Karl, 2000).
2. A context, diverse venues, and publication outlets for this debate have been provided by the US Council of Foreign Relations, a series of Washington-based foundations and policy forums as well as policy think-tanks, such

as the Carnegie Endowment for Peace and the Harvard International Center, concerned with the worldwide promotion of democracy.

3. On this see Anne Krueger (1974), former chief economist at the World Bank and currently acting manager director of the IMF, and other exponents of 'the new political economy' as well as the World Bank, which has swallowed whole and widely disseminated this 'theory' of the state. Krueger is a leading exponent of the 'new political economy,' an approach that represents at a theoretical level the neoconservative ideological offensive of capital against labor under conditions of systemic crisis. The 'new political economy' emphasizes the superiority of the world market, freed from government constraint and interference, as an engine of economic growth and development, and the private sector as the driver of this engine. Within this political economy framework (for example, Krueger, 1974; Rondinelli, McCullough and Johnson, 1989; Bardhan, 1997) the central focus of analysis has been on the propensity of governments toward rentierism, the economics of corruption, the role of governance in economic development and the economics of decentralization in 'less developed countries', as well as the need to manage the eruption of ethnic tensions and violence, i.e., the problem of 'governability.'

4. This notion, the *homo economicus* in the theoretical discourse of microeconomics, is the basis not only of neoclassical economics and neoliberal thought, but of a model of political behavior and the functioning of the state elaborated by Anne Krueger and associates—the 'new political economy.'

5. Roberto Martinez Nogueira (1991), 'Los pequeños proyectos: ¿microsoluciones para macroproblemas?', in Roberto Martinez Nogueira (ed.), *La trama solidaria.pobreza y microproyectos de desarrollo social* (Buenos Aires: GADIS, Imago-Mundi).

CHAPTER 4

1. The data presented in this chapter were collected from diverse sources, including interviews and conversations with leaders and activists of the major organizations involved in the struggle for change in Ecuador. Discussions and conversations between us and the protagonists and observers of developments in Ecuador took place in 2003 (April 7–21) and 2004 (March 7–17) with individuals and groups connected to the following organizations: Leonidas Iza, President of CONAIE; Humberto Cholango, currently president of *Ecuarani*; the leader and various 'officials' of Pachakutik; UNE—*Unión Nacional de Educadores* (National Federation of Teachers); CMS—*Coordinadora de Movimientos Sociales* (Coordinator of Social Movements); a group of union leaders of *Petroecuador*; *Universidad Central de Ecuador* (Economics and Sociology departments); the Pontífico Universidad Católica de Ecuador; the Universidad de Cuenca (Economics, Sociology and Political Science). Interviews and discussions were also held with a number of independent scholars, writers, and political analysts, including Guillermo Navarro, an independent political economist who

has published widely on diverse issues of government policy, Pablo Dávalos, a scholar and consultant close to Pachakutik, and Napoleon Saltos, a sociologist who currently heads the Sociology Department of the Universidad Central while continuing to play a leading role in the CMS.

2. According to CEPAL data, Latin America and the Caribbean are home to an indigenous population of somewhere between 33 and 40 million belonging to some 400 ethnic groups. Around 90 per cent of this indigenous population are accounted for by just five countries—Peru (27 per cent), Mexico (26 per cent), Guatemala (15 per cent), Bolivia (12 per cent), and Ecuador (8 per cent). As for Ecuador's indigenous population, estimates (CAAP, 1998) vary from 11 per cent to 30 per cent out of a total population of around 12 million (in the 2001 Census). As to 'definition' (and thus legal recognition) there are issues of language—of self and parents, residence and culture, as well as 'self-definition.' By the latter criterion only 6.6 per cent of the population would be classified as 'indigenous' (INEC/SIISE/INNFA/MBS/UNICEF/UNFPA, 2001: 89), although 14.7 per cent of those polled had parents who spoke primarily an indigenous language (17 per cent were either in this category or self-identified as indigenous). The most useful criterion for classification of the population as indigenous or not is probably residence—the question of whether or not there is a direct connection to an indigenous 'commune,' 'community,' or 'territory,' the vast majority of which are rural. As to the ethnic composition or nationality of the indigenous population, as of 1988 there has been a tacit and then explicit legal recognition of the fact that the indigenous population can be categorized into diverse 'nationalities.' The 1988 law recognized eight 'nationalities' and 14 'peoples.' Ten years later, in the decree that established CODENPE, there was another enumeration and then again in 2001, at which point the number of officially recognized indigenous 'nationalities' had increased to 13 while the number of recognized indigenous 'peoples,' a distinct category established in the 1998 law, was 18 (CODENPE, 1995: 25; Proyectos de Ley, 2001, art. 7), and thus, strictly speaking, no longer 'indigenous.'

3. Land reform in Ecuador was instituted under two military regimes, first in 1964 and again in 1971. In both cases, however, the land reform law was clearly and unambiguously tilted in favor of the big landholders (the *latifundistas*), securing their ability to work their landholdings productively and convert their traditional estates into agro-export capitalist enterprises, while insulating them from protest actions initiated by '*los indios*' who were removed from the estates and given access, if not legal title, to land that was generally poor, eroded, and non-productive (ECUARURI: 1998: 30–1, 44–7). On the 'modernization' agenda of the two military regimes implementing these land reforms, see Barsky (1986) and Santana (1984).

4. In 1972 CEDOC changed its name, suppressing the word 'Catholic,' and replacing 'workers' (*obreros*) with 'organizations' (*organizaciones*), but clearly retaining the class character of the organization which was formed in 1938 as the '*Central Ecuatoriano de Obreros Católicas*' and remained the

principal organizational vehicle for the indigenous peasant communities throughout the 1960s and the 1970s in their struggle 'for a real land reform [and] . . . the organization of agricultural day laborers,' the struggle 'against the intervention of the government in the indigenous organizations,' and 'the demand to abolish the death squads of rural police [and] the most energetic sanction of the assassins of peasants and workers' (ECUARANURI, 1998: 27).

5. CONAIE, as the greatest institutional expression of this struggle, was formed in 1986 in the coalescence of two regional organizations—ECUARUNARI, representing the land struggle in the 1950s and 1960s (viz. the government's land reform program in 1964 and 1973) of the highland Indians, and CONFENAIE, representing the struggle of Indians in the Amazonian region under the auspices of religious missions; CONAIE also involved the confluence of two major currents in the indigenous movement—a class line in the struggle against the *latifundistas* and an ethnic line concerned with the recovery and defence of 'ancestral culture.'

6. In each indigenous community there is at the same time a respect for, and deference to, the traditional authorities and leaders of the movement and democratic participation of community members in decision-making, the distribution of tasks and responsibilities, planning, political and social control (ECUARUNARI, 1998: 258).

7. During the 1990 uprising there did not yet exist the '*Coordinadora de Movimientos Sociales*' which would emerge in the mid 1990s to coordinate the protest actions of diverse urban groups and social organizations. What existed at the time was the '*Coordinadora Popular de Quito,*' a loose grouping of social organizations that responded to the call by CONAIE for fundamental change and the meeting of the basic demands of the organizations in the popular movement (Hernandez, 1998).

8. In the context of the 1990 uprising, CONAIE achieved virtual hegemony over the indigenous movement vis-à-vis the two other major national confederations of indigenous people—FENOCIN, a leftist organization of indigenous peoples, peasant producer associations, and Afro-Ecuadorians from the coastal region, and FEINE, as confederation of indigenous nationalities with its strength in the Amazonian region.

9. There were two other, more distant and less immediate elements of this context. One was the constitution, on October 12, 1992, of the '*encuentro de dos mundos,*' marking the final celebration of a worldwide campaign, begun in 1990, to protest 500 years of European 'discovery' of the 'new world' and the conquest, exploitation, and continuing oppression of indigenous peoples. Another was the extraordinary assembly of the Organization of the Indigenous Peoples of Pastaza on March 2–3 1992, to protest the government's failure to legalize ownership and control by the Huar, Quicha, Shwar, and Záparo peoples of their national territory, and the 'March for Territories and Life' in April.

10. The agrarian modernization law proposed by the government was along the lines of similar legislation advanced in Mexico and elsewhere in Latin America at the time. On the measures included in this legislation, and their political significance vis-à-vis the process of 'market-assisted' (and

World Bank-designed) land reforms, see De Walt and Ress (1994) and Deininger (1998). As in a number of other countries in the region, the government-assisted land reform program of the 1960s (expropriation of large landholdings and their distribution to indigenous communities or families) had collapsed.

11. The end of Latin America's land reform program had been announced some years back, but the agrarian modernization law adopted by Brazil, Ecuador, and Mexico, among others early in the 1990s, reflected a change in the approach of governments in the region to the issue of 'land reform.' In effect, it represented a turn to an approach advocated by the World Bank, namely a 'market-assisted' approach. On the dynamics of this shift in the nature of the land reform program in Latin America, see Petras and Veltmeyer (2003).

12. The *Political Project* (CONAIE, 1994), published by CONAIE's *Consejo de Gobierno* and approved by its Congress IV, details the movement's ideas for state reform based on the vision of a multi-ethnic, pluri-national state.

13. Pachakutik is essentially an electoral apparatus, and as such the political wing of CONAIE. However, its protagonists have always rejected calling Pachakutik a political party, viewing it instead as a political movement with a relative autonomy vis-à-vis CONAIE.

14. By all accounts the process of consolidating Pachakutik as a political movement and electoral instrument was complicated and fraught with difficulties and setbacks, not least of which was the continuing divisions between the highland communities organized under ECUARUNARI and the Confederation of Amazonian nationalities. The former conceived of the need to build a political movement (the *Movimiento de Unidad Plurinacional Nuevo País*) but to do so by degrees, building from the bottom up, while the latter was concerned to establish a political instrument (Pachakutik) that would enable the indigenous communities to participate immediately at all levels of electoral politics, and to contest state power directly, in all of its forms, including the presidency. An agreement was finally reached in 1995, allowing CONAIE to consolidate itself as both a social and a political movement, albeit not without internal contradictions and divisions.

15. This and other such electoral alliances have their limitations and pitfalls, as demonstrated by what is now regarded by the leaders of CONAIE as its disastrous alliance with Lucio Gutiérrez's PPP. In the case of Freddy Ehlers, the official candidate of the CMS in the 1996 elections, the problem was that in the interest of establishing a political base within the bourgeoisie he had a very different idea from CONAIE as to the strategic areas of the economy and the need to retain and strengthen the public sector in these areas. This program incorporated the basic demands of CONAIE (and Pachakutik), namely the constitution of Ecuador as a pluri-national, multicultural state; and the constitution of an indigenous parliament and popular assemblies at all levels from the local to the provincial. Ehler was prepared to turn over the strategic sectors of oil, electric power, and telecommunications to foreign capital, an issue on which he would never had achieved an agreement with CONAIE, the leadership of which saw

these areas as 'strategic areas' of the economy to be protected at all cost and maintained in the public sector in the interest of all Ecuadorians.

16. Lucas (2003) in this connection writes about and warns of a division between the social movement, represented by CONAIE, and the political movement represented by Pachakutik—and its political project—to participate in the electoral political system of the state. As Lucas constructs it, this division reflects the dilemma of a movement caught between a class-oriented old left and a new left oriented toward cultural issues of political identity. The issue: should the Pachakutik movement transform itself into another center-left political party with its political base in the indigenous community and the urban poor, and with the support of social organizations and NGOs of Ecuador's civil society, contest national and local elections? In effect, should the indigenous movement seek to transform the Ecuadorian uni-national state into a pluri-national one from within or via the agency of an independent social movement?

17. In the 1990s 'good governance,' a concept originated by the World Bank in its 1989 report on Sub-Saharan Africa, became a critical component of its program of neoliberal reforms. In Spanish, the translation of 'governance,' which, unlike 'governability,' implies a form of political order (on the basis of social consensus and with as little 'government' as possible) rather than the capacity to govern/establish political order, is *'gobernabilidad.'* Strictly speaking, this translation conflates the notion of 'governance' in its neoliberal sense and 'governability' in the meaning assigned to it by Crozier, Huntington, and Watanuki (1975) in their report to the Trilateral Commission concerning the political impact of excessive democracy (too much participation placing undue pressure on the limited capacity to govern and maintain order).

18. Only three countries (Bolivia, Peru, Venezuela) experienced a worse deterioration in the purchasing power capacity of incomes and wages than Ecuador over the course of two decades of structural adjustment. With an index of this capacity set at 100 in 1980, in 1998 it went up to 190 for Latin America as a whole while in Ecuador it fell to 85—60.2 in Bolivia and 59 in Venezuela (Larrea, 2004: 79, with CEPAL data).

19. Ecuador's program of structural reform and convertibility plan can be usefully compared to the similar measures taken in Argentina, albeit in a different *de facto* form (constitutionally legislated parity), as well as Ecuador's legislated dollarization in 2000.

20. One way Bucaram's political enemies did this was to use a congressional approval of a highly unpopular *'paquetazo'* of macroeconomic policies that the administration in fact sought to avoid. Precipitous congressional action forced Bucaram's hand, placing his government in a difficult position—between a rock and a hard place, as it were.

21. On the resulting micro-dynamics, see Cameron (2003).

22. SIOS (2002) has identified close to a 1,000 NGOs operating in Ecuador, most of them formed in the 1980s and the 1990s. On the role of these NGOs in Ecuador's local development process, see Bretón (2003).

23. These payments on the external debt were made in the context of a banking crisis whose proportions equalled or exceeded the worst of many such crises in the region, the banking crisis in 1983 which ended the first

round of neoliberal experiments initiated by the Pinochet regime in Chile, a crisis that entailed a bailout equivalent to 6 per cent of Chile's GNP at the time. In the case of Ecuador, the failure of bank after bank from 1995 to 1999 was the direct result of the government's financial liberalization measures (Salgado, 2000). A bailout, in 1999, of the banks that held 41 per cent of all private external debt in the country cost the public purse more than $3.5 billion, and if one were to add to this the devastating impact of the government's policy of freezing bank deposits as well as the issuing of bonds to capitalize several banks, policies that in Argentina several years later would have such a devastating economic and political effects, then the scale of Ecuador's financial crisis was such as to push the whole economy into the country's worst crisis to date. The banking crisis of 1999 was the result of the confluence of the government's policy of financial liberalization and the economic crisis. The government's response was to bail out the biggest banks, converting what was a privately held debt into a public debt, and a policy of dollarization, to relieve thereby intolerable pressures on the *sucre*'s exchange rate as well as the rates of inflation and interest, which in March soared to 150 per cent (Salgado, 2000: 12–13). On the economic and political dynamics of this financial crisis and the subsequent dollarization of the economy see, *inter alia*, Salgado (2000).

24. The IMF's letters of intent signed by the government included a commitment to push ahead and implement the following policies: (1) the adoption of 'realistic' exchange rates and prices (periodically raised) for 'public goods' related to transportation and basic needs; (2) an increase in the tariffs for water, electricity, telephone services, etc; (3) a reduction of public sector employment and the closing of government offices; (4) an increase in the bands of tax payments; (5) the privatization of state enterprises, beginning with Ecuadorian airlines, *Ingenious Azucarerero* AZTRA and FERISA, but moving on to EMETEL, INECEL, Correos and, most significantly, PETROECUADOR (the state oil company) and IESS (public social security); (6) liberalization of the banking sector and deregulation of private finance; (7) negotiations with foreign banks and creditors in regard to the public debt; (8) an opening toward and removing restrictions on the entry of foreign direct investment; (9) eliminating the policy of automatic adjustments such as the indexation of salaries and wages; (10) eliminating government interference in the labor market, allowing for the 'free' exchange of capital for wages—the contracting of labor; and (11) increasing non-traditional exports (Vicuña Izquierdo, 2000: 6–7).

25. Another dimension of this decapitalization is provided by what has been regarded as the 'biggest swindle of the twentieth century,' namely the bailout of the country's private banks to the tune of $4 billion, the socialization of this debt and the privatization of the bailed-out banks (Vicuña Izquierdo, 2000: 176). In regard to the bailout of the banks after the 1998 financial crisis, Wilma Salgado (2000) estimates that its total value was equivalent to one third of Ecuador's GDP in 1998, making it, as she points out, one of the most costly bailouts in Latin America's history of major bailouts dating from the bailout of the private banks by Chile's military regime in 1983, a bailout equivalent to 6 per cent of Chile's GDP

at the time. On the connection between the bailout of the banks and the subsequent dollarization policy, see Delgado Jara (2000).

26. By some, if not all, accounts this historic event (January 1, 2000) although traceable to a long historic struggle against Creole domination, was triggered by a policy decision announced some ten days earlier by Mahuad to adopt the US dollar as the national currency. This policy implied an exchange rate of 25,000 *sucres* to the dollar in a country in which the minimum wage was barely $53, the cost of a monthly basket of basic goods was $200 and over 70 per cent of the population—over 90 per cent of the indigenous population—lived below the World Bank's conservatively defined poverty line.

27. Accounts of the 1990 uprising were dominated by anthropologists, sociologists, and historians. In contrast, accounts and analysis of the January 21 uprising involved not only the usual spectrum of social scientists but also politicians, journalists, and military officials, each seeking to document aspects of the political dynamics involved. For a compilation of press accounts one can consult various sources, but most 'accounts' are descriptive rather than analytical. Even Saltos (2001), a sociologist with a good grasp of the political dynamics of the uprising, analyzes events from the perspective of a participant, in this case as a leading member of the CMS which joined with CONAIE to orchestrate political actions in a quasi-revolutionary situation. An overall analysis of January 21 that takes account of diverse, if not all, conditions and agencies, has yet to be written—with scholarly hindsight.

28. By diverse accounts there were at least four conditions of this situation: (1) opposition to the government's measures for settling the border issue with Peru; (2) reduction of the budget for the armed forces, with its negative salary and spending implications; (3) perceived corruption and widespread anger at the government's dealings with, and bailout, of the banks (this issue figured most prominently in the political discourse of the military officers that joined the uprising); (4) the suggestion of an agreement between Mahuad and the *Partido Roldista Ecuatoriano* (PRE), the party of the deposed ex-president Abdalá Bucaram, that included (apparently) his return from exile in Panama; and, in addition (5) it appears that informal connections and conversations between elements within the armed forces and the indigenous movement existed as of the 1990 uprising and increased over the years. It is clear that the high command of the armed forces were by no means disposed to a military uprising or coup. At the same time, it was not in a position, or able, to violently oppress a movement in this direction, not even within its own ranks.

29. CONAIE had come to an agreement on 23 points with the Noboa government on February 7, 2001. This agreement, signed by Noboa himself, included a commitment of the government not to proceed with its IMF-mandated austerity program and to advance a program of economic and political reforms, including the decentralization of the government.

30. On the 'contradiction' or strategic choice implied by the political and social division of the indigenous movement see, *inter alia*, Kintto Lucas

(2003: 78, 81–3). One of the issues involved in this division (in addition to the question of how best to achieve state power) is how to conceive of Pachakutik—as the 'political arm' of CONAIE, and thus part of and accountable to it, or as its 'progeny' and thus to some extent 'autonomous' or 'independent' from it. It seems that this issue has been the subject of considerable internal debates within ECUARANI and other organizations that make up CONAIE and constitute Pachakutik.

31. The one leftist candidate in the elections, presented by the MPD, resigned in support of Colonel Lucio Gutiérrez who, for some inexplicable reason, was regarded (or at least defined) by the MPD as a leftist (in *Tintají*, No. 9, p. 8).

32. Also, when asked if his '*Sociedad Patriótica*' was a leftist organization, Gutiérrez responded: 'No. If you want to place me on some political line it would have to be in the center' (*El Universo*, January 2, 2002).

33. At the time of writing, Gutiérrez's party had not yet informed the TSE about the financial contributors to its campaign, despite the law obliging it to do so. It is clear that between the first and second rounds of the presidential vote numerous contributions came Gutiérrez's way, undoubtedly even from the oligarchy against which his initial campaign rhetoric was directed.

34. This was confirmed by the leader of Pachakutik himself in an interview with the authors in October 2002, less than a month before the elections. It was here confirmed, in effect, that neither CONAIE nor Pachakutik had formulated a government program or placed any programmatic demands on Gutierrez—not even a commitment to some form of land reform or opposition to ALCA, an issue of political principle for all forces of political opposition to the government of the day.

35. The letter of intent required the adoption of a neoliberal program of policy measures that included a commitment (1) not to incur any further arrears in debt repayment; (2) not to impose any restrictions on international trade (ALCA, etc.); (3) to maintain the US dollar as the national currency; (4) to effect a 35 per cent increase in the price of fuel to generate US$ 400 million (1.5 per cent of GNP); (5) to end the subsidy on gasoline; (6) to privatize the Banco del Pacífico; (7) to ensure the administration of the telephone and electrical utilities by foreign companies; (8) the freezing of pensions; and (9) the implementation of fiscal, tax, and labor reforms (implying an increase of taxes on workers with low and medium incomes). The budget proposal sent by the government to the IMF allocates 35.7 per cent to debt service versus 19.5 per cent for social programs in the areas of health, education, housing and welfare. The share of education in this budget would be 11.2 per cent (versus 12.5 per cent in 2002).

36. In regard to these attacks, of course, Gutiérrez is honoring what is now a tradition among governments in Latin America that have taken their cue from the World Bank in its double-edged campaign to convince workers that globalization and structural adjustment is in their interest (World Bank, 1995); and blame organized labor and government interference in labor markets as the cause of Latin America's 'ills'—poverty, informalization, unemployment, social exclusion.

37. This world is not only culturally diverse but organizationally very complex. In fact, the term 'indigenous community' encompasses but tends to obscure the existence of multiple forms of 'base organizations' (communities, cooperatives, water boards, associations, women's groups, clubs, religious and other groups) and the term 'community,' connoting relations of solidarity and a sense of identity and belonging that can reach beyond a specific 'territory,' can be distinguished from 'commune,' denoting more of a political or administrative grouping (smaller than a *parroquia*). On these and other considerations, see the most useful discussion by Guerrero Cazar and Ospino Peralta (2003: 30–1).

38. The hegemony of CONAIE within the movement has not been uncontested. Contestation of this hegemony has come from within the movement and from other indigenous organizations such as FENOCIN, a federation of indigenous peasant farmers and Afro-Ecuadorians associated with the Socialist Party.

39. On April 28, 2004 CONAIE announced (see Argentinpress.info) a mobilization to force the 'traitor' from office, the only possible and necessary means of resisting his 'neoliberal and anti-popular' policies. It also objected to and rejected the repeated interventions of Washington's ambassador (Kristie Kenney) to Quito, including her efforts to approach community leaders in an effort to co-opt them. At the same time; however, Pachakutik announced its intention to participate in the October elections with its own candidates, indicating thereby that the indigenous movement had not given up on electoral politics as a path toward power.

40. The World Bank and other development agencies for International Cooperation generally do not work with class- or community-based grassroots organizations (CBOs) but prefer—in fact insist on—working through or, if necessary, setting up for the purpose, 'second grade organizations' (OSG—*Organizaciones de Segundo Grado*) deemed to have greater 'institutional capacity' and a broader base of 'stakeholders.' The base organizations affiliated with these OSG assume a variety of forms (*comunas*, cooperatives, *asociaciones*, etc.), whose dynamics regarding their relations to the 'community' on the one hand, and outside organizations on the other, are surprisingly complex (Chiriboga, 1985; Zamosc, 1995). Regarding the intermediary position and role of the nongovernmental organizations that have invaded the Ecuadorian countryside over the past decade, and the relationship between these NGOS and the OSG and the CBOs, there are two fundamentally opposed theoretical (and political) perspectives. A contrary—that is, critical—view is very well argued in an analytically profound study by Bretón (2001).

41. This view of PRODEPINE is widely held among the indigenous intelligentsia and community leaders. On this issue see in particular the opinions disseminated through the *Bolétin ICCI-RIMAY* (2001) of the Scientific Institute of Indigenous Culture (ICCI). Textually in this document the dominant opinion presented is that '[t]he enormous resources possessed by this organization (PRODEPINE) constitutes a permanent threat to the indigenous movement. The modernizing vision and neoliberal parameters used by the technobureacrats of this institution

constitute a threat to the political project of the indigenous organizations and a source of permanent conflict.' The text adds that 'it is not a matter of a clash between two distinct visions of reality [Huntington's "Clash of Civilizations"] but of a fundamental confrontation between two distinct historical [political] projects.'

42. The Shuaras, a major indigenous nation in Ecuador's Amazonia, have noted that unlike the highland Quichuas they had never been conquered but, at the same, they have 'lived a strong process of colonialism as regards the church, oil companies and innumerable NGOs' ('Antecedentes al surgimiento de pachakutik, *Riccharishun*, December 1, 1998).

43. In this connection the 'alternative local development/democratic governance' strategy pursued by the World Bank and other organizations of 'international development assistance' is proving to be more effective than the programs of integrated rural development pursued to the same purpose (turn the indigenous movement toward a confrontationalist politics and a program of social and political reforms) of the 1970s.

CHAPTER 5

1. In an interview with *Punto Final* (May 2003: 16–17) Evo Morales, leader of the major political force on Bolivia's left, the *Movimiento al Socialismo-Instrumento Politico para la Soberanía de los Pueblos* (MAS-IPSP), defined socialism in terms of 'communitarianism.' This is, he notes, because 'in the *aylla* [the principal *Aymara* territorial unit] people live in community, with values such as solidarity and reciprocity.' 'This,' he adds, ' is our [political] practice.'

2. It is estimated (by Morales, 2003) that 60 per cent of the total population is indigenous and belong to the dominant original peoples of Bolivia, the *Aymara* and the *Quechua*; and from 80 to 90 per cent of the *altiplano* peasants have parents belonging to these two indigenous nationalities. In addition, official statistics suggest that a smaller but growing proportion of the population, dispersed across the Amazon, the eastern lowlands and the Chaco region, belong to some 32 ethno-culturally distinct groups such as the Guareníe and Chiquitanos, who collectively make up less than 3 per cent of the national population.

3. On these conditions see, *inter alia*, INE (2002).

4. The agrarian reform strengthened the peasant movement in the *altiplano*, and, with the encouragement of the MNR, who had its own reasons for doing so (see the discussion below), the first national peasant union was created—the National Confederation of Peasant Workers of Bolivia (CNTCB) to which peasants were automatically affiliated at the community level (Albó, 2002: 75). In 1971, after years of a military-peasant pact forged by the MNR as a weapon to be used against the powerful mining unions, a movement for the independence of the *Aymara* peasantry from the government led to the formation of the Union Confederation of Peasant Workers of Bolivia (CSUTCB) in 1979.

5. The CSTUCB joined the COB and several left-wing political parties to overthrow two violent, albeit short-lived, neo-fascist dictatorships,

formed in a military coup led by Hugo Banzer Suárez, on the basis of a military–peasant pact, against the Popular Assembly government of José Torres from 1969 to 1971. The overthrow of the dictatorship gave rise to the election of an equally short-lived center-left UDP government in 1982. The inability of this popular, democratically elected government to 'govern' under conditions of runaway inflation, economic sabotage, and political opposition that resembled the situation faced by Salvador Allende in Chile (and Chavez in Venezuela) led to the formation of an MNR-led coalition government and neoliberal regime in 1985.

6. The legal institutionality of this model was established by presidential decree rather than congressional legislation—Decree 21060 (1985). This decree, which, among other measures, included the closure of the tin mines and the 'relocation' (firing) of the 10,000 miners who formed the backbone of the economy at the time, was supplemented and modified by subsequent 'supreme presidential' decrees in 1987 (2DS 21660) and 2000 (DS 22467). Together, these executive decrees constitute the legal-institutional foundation of the macroeconomic policies implemented over the past two decades on the basis of the 'new economic model' (neoliberalism).

7. On this battle, see Kruse (2002).

8. The MIP should be clearly distinguished from another current in *Aymaran* politics, *Katarismo* or MRTK, represented by De Lozada's running mate and vice-president in his first administration. This current was also rooted in the ideology of '*Aymaran* nationalism,' but it has been much more integrationist, in favor of assimilation into the Bolivian state. The MIP, in contrast, is opposed to both this state and the '*k'ara blancos extranjeros*.' At the same time, MIP is profoundly indigenist and suspicious of the traditional left who, in their *k'aras blancos* and politics, are generally excluded from its politics much as '*los blancos*' (most of whom are in fact *mestizo*) who dominate the political left in Peru are seen as part of their enemy by *senderismo luminoso*.

9. For an exposition of this ideology, see Quispe Huanca (1999).

10. On this form of development, based on the building of social networks rather than political power, see the discussion below.

11. Unlike Quispe, who has more or less stuck with the social movement against Mesa's cooptation tactics, Durán succumbed to pressure for the MST to bring an end to the land occupations and 'resolve the conflict through dialogue . . . within a legal framework and respect for private property' (Jorge Cortés, Minister of Sustainable Development, *argenpress. info*, November 13, 2003).

12. Although this point has been disputed. Some observers (Econoticias/ Argenpress.info. November 13, 2003) identify as the leading forces in the October Rebellion those led by Quispe together with COB, the regional Workers Central and the Federation of Neighborhood Associations in El Alto.

13. The population of El Alto represents 28 per cent of the total population in the Department of La Paz (INE, 2002). 81.3 per cent of this population defines itself as 'indigenous,' primarily *Aymara* (an indigenous group that makes up about a quarter of Bolivia's population of 8.5 million)

and, according to official statistics, 51 per cent of the Department La Paz is classified as poor, unable to meet their basic needs, with the greatest concentration of poverty in El Alto. Given this 'poverty' a large part of the indigenous population in El Alto is found in what sociologists classify as the 'popular sector,' but it also meets the classical definition of a 'proletariat,' namely, a population dispossessed of any means of production.

14. The 'system,' formed in 1985, is composed of three parties at the center (MNR, MIR, and ADN) that have accounted for around 65 per cent of the national vote in most elections over the next two decades, and four on the margins (MBL, UCC, Condepa, and NFR), with which the core parties formed coalitions and shared power in different combinations as required or served as a 'functional opposition,' also as required for maintaining a democratic façade (*30 Días*, February 2002, p. 65).

15. The direct and public intervention of US ambassador Rocha in these elections was controversial to say the least, but deemed necessary given the perceived threat represented by Morales, who was publicly denounced and demonized by the ambassador, who openly recommended that the electorate not support his candidacy and that of MAS (see *30 Días* July 2002, p. 7 on the furore caused by this precipitous 'intervention' in Bolivia's political affairs by the leading representative of, in Rocha's words, 'one of the biggest and most important' embassies on the continent, see *30 Días*, July 2002, p. 17). Of course, such US intervention is not new; in fact it is all too common. But rarely is it this overt.

16. Over 20 years of 'democracy' in Bolivia, MIR has shared power a number of times: in 1985 MIR turned its 15 congressional votes over to Paz Estenssoro, allowing him to become president; in 1989, Jaime Paz, the leader of MIR, assumed the presidency with the support of Banzer (UDN); and in 2002 MIR entered into a co-government arrangement with De Lozada–Mesa (MNR). That MIR, despite its ideology, is also part of the system is evidenced not only by its willingness to join any government in power in exchange for a negotiated share of government positions, but by the class composition of its political representatives, many of whom are very much a part of the small minority elite, an oligarchy of a small number of political families. In the case of the MNR-MIR alliance just seven family groups control both the executive and legislative branch of government (*30 Días*, August 2002, p. 15).

17. Just two weeks before the elections the polls gave the NFR, represented by Manfred Reyes Villa, ex-governor of Cochabamba, 27 per cent of the vote, which might very well have given him the presidency. However, as it turns out, he received only 20 per cent of the vote, having been beaten soundly in Cochabamba, the bastion of his political support as well as locus of the water war, by MAS, which had joined the *Coordinara* against the privatization and export of water. In the week before the actual election the US ambassador, Manuel Rocha, publicly intervened in the electoral process by denouncing Morales and recommending voting against him (at the cost of losing US government aid), a factor that probably consolidated a first place finish for Morales and MAS in Cochabamba.

18. This march, and the associated movement to create a more participatory state, was joined by other organizations including the *Confederación de Colonizadores* (in the *east*), *the Federación de Mujeres Campesinas Bartolina Sisa*, the *Coordinación de Integración*, the *Movimiento sin Tierra* (MST), and a number of other peasant federations and social groups.

19. This *'perdonazo'* of debts via the Law of Enterprise Restructuring contrasted markedly with the refusal of the government to support the thousands of small businesses and producers condemned to bankruptcy by crippling debts arising out of commercial bank rates exceeding 35 per cent (Coggiola, 2003: 30).

20. In less than a month, mostly from October 9 to 18, ten days that 'shook Bolivia' and the world, troops under the command of the MNR killed more than 84 civilians and 15 conscripts who refused to fire on unarmed protesters. A further 40 or so people were 'disappeared,' over 500 seriously injured and an untold number detained in a desperate effort by the government to maintain power and preserve the neoliberal status quo.

21. MAS, without a doubt, has considerable political influence in a number of social organizations in the popular movement, including in particular the *Confederaciones* of Rural Teachers of Bolivia and La Paz, the Regional Workers Central of El Alto (led by Juan Melendres, who is part of MAS's central committee), the Federation of Neighborhood Associations in El Alto (led by another MAS militant, Mauricio Cori), the *Confederación de Colonizadores de Bolivia*, the *Cocaleros of los Yungas*; and the peasant (Departmental) federations of Oruro, Cochabamba, Chuquisaca, Tarija, Potosí, Pando, which formed an organization, led by Román Loayza, that paralleled the CSTCUB, led by Felipe Quispe's MIP (*Pachakuti*). MAS also has influence with the MST (*Movimiento de los Sin Tierra*), *los Desocupados*, the National Association of Mining Cooperatives (*Asociación Nacional de Cooperativas Mineras*), the University of El Alto and UMSA, and intermediate Workers' Centrals in Oruro, Cochabamba, Chuquisaca, Tarija, Potosí, and Pando, the *Confederación de Fabriles de Bolivia*, and a sector of *gremios* in Bolivia and La Paz. Also as of October various neighborhood association in La Paz, Cochabamba, Oruro, Potosí, Chuquisaca, and Santa Cruz joined the MAS electoral bandwagon.

22. In general terms this model prescribed policies of economic and political reform—privatization, deregulation, liberalization, decentralization, municipalization. A number of studies have shown that the Bolivian model of these policies was first tested in Bolivia and then exported for use in Central America, Eastern Europe and Africa (Graham, 1992; Brada and Graham, 1997; Peirce, 1997; Xue, 1997).

23. The innovation of the *Plan de Todos* was that it simultaneously attempted to reconcile the demands of subnational regions for greater autonomy with those of international institutions for open markets. The *Plan* allowed regions to gain some degree of autonomy and financial resources to embark on local projects through the Law of Popular Participation, while multinational firms would gain access to Bolivia's natural resources through the Law of Capitalization and related economic policies.

24. Most of the debate and discussion on decentralization and increased municipal autonomy in developing countries totally ignores the global

and political context in which these reforms take place and presents the policy of decentralization as the work of apolitical professionals operating within international financial and development institutions and see it as a matter of increased economic efficiency as well as democracy (Rondinelli, Nellis and Cheema, 1983; Rondinelli, McCullough and Johnson, 1989; World Bank, 1997; Grindle, 2000; Van Cott, 2000).

25. On the current dominant development paradigm and these two agendas, see World Bank (1997).

26. Promoting democracy as a way to guarantee political stability for economic globalization is central to the neoliberal model. Just as the World Bank's program of Social Emergency Funds under the 'New Social Policy' was first pioneered and 'tested' in Bolivia during the 1985–9 MNR administration (Graham, 1992), the innovative neoliberal policies introduced in De Lozada's *Plan de Todos* (1993) were adapted for use in Central America, Eastern Europe, Asia, and Africa (Brada and Graham, 1997; Peirce, 1997; Xue, 1997).

27. Key aspects of the Law of Popular Participation (LPP) include: municipalization—the creation of new municipalities in rural areas; a doubling of the municipal share of central government revenues from 10 per cent to 20 per cent, and its allocation to each municipality on a per capita basis; and municipal title to all local infrastructure related to health, education, roads, culture, etc., together with the responsibility to maintain and improve it. Prior to the LPP, 92 per cent of state transfers were distributed to the capital cities; by 1997, according to De la Fuente (2001) they received only 39 per cent. Bolivia's municipalities, more like counties than cities, are predominantly small and rural—94 per cent of the country's 314 municipalities have populations under 50,000.

28. In the context of developments leading up to October, *El Nuevo Día* (Santa Cruz, June 15, 2003) reported on the isolation of MAS from the popular movement in its electoral strategy, supported by NGO 'professional' advisors, and concern to win over the disaffected sectors of the middle class as well as the *gremios* in sectors such as transportation. In this strategy MAS was clearly isolated from the growing popular movement and its strategy of mass mobilization. It is safe to say that in its politics, MAS has distanced itself from the popular movement.

29. An example of this approach can be found in the massive marches and demonstrations in La Paz on May 24, 2004. In this mobilization thousands of people pressed for the renationalization of gas and protested the government's proposed referendum. 'The people,' it was argued, 'have already spoken on the issue.' In La Paz and El Alto, 83 per cent of the people, according to a radio poll by Erbol, are in favor of nationalization; only 13 per cent were against. At the same time, 69 per cent declared themselves in favor of exporting gas via Peru rather than Chile, a position that clearly reflects the troubled history of Chile's denying Bolivia access to the sea in the war of 1879–84.

30. With the privatization of pension funds in 1997, and a capital flight that exceeded $300 million, the deficit on the government's capital account grew from 321.1 million Bolivianos in 1997 to 1.9 billion in 2002 (Coggiola, 2003: 31). As for the AFPs, the mechanism set up by the

government to hold the proceeds of capitalization (50 per cent of the shares of the privatized firms), total proceeds of this program amounted to only 4.8 per cent of government revenue, or 0.66 per cent of GNP, not nearly enough to cover the huge and growing external debt. In this context of growing indebtedness, the government negotiated a royalty share of only 18 per cent of the revenues derived from the sale of gas and oil versus an industry average royalty share of over 30 per cent.

CHAPTER 6

1. The World Bank, the Inter-American Bank, and other such organizations involved in the process of international cooperation for development, a project that can be traced back to the post-Second World War geopolitical concern that countries might be tempted down a socialist path of national development, have elaborated variations of an approach, a model of development based on the accumulation of 'social capital'. The literature on this approach to social change and economic development is voluminous, most of it supportive. For a critical perspective on this model and the 'social capital' approach to development (micro-projects) and politics (local democracy and governance), see, *inter alia*, Harris (2001), and Schuller (2000).

2. Harris (2001) is one of few authors to provide a critical perspective on the World Bank's construction of this concept of 'social capital'. See also Veltmeyer (2002).

3. The creation of a parallel organization typically involves staged elections for a new board of directors. Government agencies or the courts then award the organization's legal identity (along with offices, bank accounts, and other resources) to a favored faction, whether or not it represents the membership.

4. Governments in the region have frequently resorted to repression as a means of demobilizing organizations with an anti-systemic agenda. At times, it has involved the full weight of the state's repressive apparatus as in the dirty war orchestrated by a coalition of armed forces and a series of authoritarian-bureaucratic or military regimes in the Southern Cone of South America against the labor movement in the 1970s. In other conjunctures, as in Ecuador in the mid-1980s, the instruments of state terror and repression were wielded against the working class by regimes that are formally democratic. In this conjuncture—and other such conjunctures in the 1980s in Bolivia, Venezuela, and elsewhere, involving conditions of a brutal repression—radical opposition to the government's neoliberal agenda, led at the time by the labor movement, was disarticulated and demobilized, weakening and close to destroying working-class political organizations in the process (Editorial, *Boletín ICCI*, Vol. 1, No. 8, November 1999). As it happens, in the case of Ecuador, the repression and destruction of the labor movement's capacity to challenge the government's agenda coincided with the emergence and formation of CONAIE which, in the 1990s, would take over leadership of the popular struggle.

5. In the context of conditions found throughout the region in the 1990s, a marked development and trend was toward the disarticulation of class-based organizations and a demobilization of the forces that they had accumulated and mobilized. The dynamics of this political demobilization are not well studied or understood, and there are doubtless many factors involved. However, it is also doubtless the case that a combination of strategies pursued and implemented by governments in the region, and with the support of both outside or international organizations and NGOs within, was a critical factor in the widespread demobilization of many social movements in the 1990s. This factor is clearly evident in the case of the *Alianza Democratica de Campesinos* (ADC), which, in the post-civil war context of El Salvador, emerged as the most representative and dynamic social movement of peasants organized around the issues of land redistribution and indebtedness. As a coalition of diverse peasant organizations, the ADC initially pushed its land reform agenda through a politics of direct action (land invasions, marches, and so forth) but was soon constrained to operate within the framework of reforms established through the peace accords. Under these conditions, and with the active support of the *Frente Farabundo Marti de Liberación Nacional* (FMLN), transformed from a belligerent armed force into a leftwing political party, the struggle for cancellation of the land and bank debts was more or less resolved in political-legal terms (through legislation) in the interest of the beneficiaries of the first phase of the government's land reform program. However, all direct and even indirect action on the land issue was definitively stalled by a politics of economic development projects funded by the World Bank and other donor agencies and executed through NGOs. On some dynamics of this process, see Veltmeyer (1999).

6. This strategy was pursued and implemented by all multilateral and bilateral development agencies in the 1990s.

7. Various analysts are of the opinion that Morales' electoral approach toward politics was influenced by his November 2003 meeting with Brazilian PT President Luiz Inácio (Lula) da Silva, who lost three bids for the presidency before achieving power in 2002. Needless to say, Lula told Morales to be patient, learn from his defeats so as to represent all Bolivians—even the economic elite. As Lula put it: 'you cannot be limited to being an indigenous leader or a former leader.' He did not have to add that electoral politics is the only way to state power.

Bibliography

Acosta, Alberto and Lautaro Ojeda (1993). *Privatización*. Quito: Centro de Educación Popular.

Albó, X. (2002). 'Indigenous Political Participation in Bolivia,' in R. Sieder (ed.) *Multiculturalism in Latin America: Indigenous Rights, Diversity and Democracy*. Basingstoke: Palgrave Macmillan.

Annan, Kofi (1998). 'The Quiet Revolution,' *Global Governance*, 4(2): 123–38.

Arbos, Xavier and Salvador Givier (1993). *La gobernabilidad, cuidadania y democracia en la encrucijada mundial*. Barcelona: Editorial Siglo XXI.

Ardaya, R. (1995). *La construcción municipal de Bolivia*. La Paz: Strategies for International Development.

Arias Duran, I. (1996). *El proceso social de la participación popular: problemas y potencialidades*. La Paz: SNPP.

Assies, Willem (2003). 'David versus Goliath in Cochbamba: Water Rights, Neoliberalism and the Revival of Social Protest in Bolivia,' *Latin American Perspectives*, 30(93): 14–36.

Ayo, D., C. Barragan and O. Guzman Boutier (1998). *Participación popular: una evaluación aprendizaje de la ley 1994–1997*. La Paz: Ministerio de Desarrollo Sostenible y Planificación; Viceministerio de Participación Popular Fortalecimiento Municipal, Unidad de de Investigación y Análisis.

Bardhan, Pranab (1997). *The Role of Governance in Economic Development*. Paris: OECD, Development Center.

Barsky, Osvaldo (1986). *La reforma agraria ecuatoriana*. Quito: CEN.

Bebbington, Anthony (2001). 'Development Alternatives: Practice, Dilemmas and Theory,' *Area*, 33(1), pp. 7–17.

Benitez, Milton (1992). 'Perfiles de la democracia ecuatoriana,' *Ecuador en la postguerra*, Banco del Ecuador.

Besayag, Miguel y Diego Sztulwark (2000). *Política y situación: de la potencia al contrapoder*. Buenos Aires: Ed. De mano en Mano.

Bhagmati, J. (1995), 'The New Thinking on Development,' *Journal of Democracy*, 6(4): 50–64.

BID (Banco Interamericano de Desarrollo) (1996). *Modernización del estado y fortalecimiento de la sociedad civil*. Washington, DC: BID.

BID (Banco Interamericano de Desarrollo) (2000). *Desarrollo: más allá de la economía. Progreso económico y social de la América Latina*. Washington, DC: BID.

Birdsall, Nancy (1998). 'Life is Unfair: Inequality in the Global Economy,' *Foreign Policy*, 111: 76–93.

Blair, H. (1995). 'Assessing Democratic Decentralization,' A CDIE Concept Paper. Washington DC: USAID.

Blair, H. (1997). 'Democratic Local Governance in Bolivia,' CDIE Impact Evaluation, No. 3. Washington, DC: USAID.

Bolivar Castillo, José (2004). 'Descentralización: Desafío de la democracia y el desarrollo,' *Tendencia. Revista Ideológico Político*, I, Quito, March, pp. 98–105.

Bolivia, Gobierno de (1994). *Ley No. 1551 de Participación Popular.* La Paz.

Bolivia, Ministerio de Desarrollo Sustenible y Medio Ambiente (1994). *Plan General de Desarrollo Economico y Social: El Cambio para Todos.* La Paz Ed.

Bombarolo, Félix, Luis Coscio Perez and Alfredo Stein (1990). *El rol de las ONGs de desarrollo en América Latina y el Caríbe.* Buenos Aires: Ficong.

Booth, David (1996). 'Popular Participation, Democracy, the State in Rural Bolivia,' Department of Anthropology, Stockholm University, Sweden.

Booth, D., S. Clisby and C. Widmark (1995). 'Empowering the Poor through Institutional Reform: An Initial Appraisal of the Bolivian Experience,' Working Paper 32, Department of Anthropology, Stockholm University, Sweden.

Brada, J. C. and C. Graham (eds.) (1997). *The Deepening of Market-Based Reform: Bolivia's Capitalization Program.* Washington, DC: The Woodrow Wilson Center.

Bretón Solode Zaldivar, Victor (2003). 'The Contradictions of Rural Development NGOs: The Trajectory of the FEPP in Chimborazo,' in *Rural Progress, Rural Decay: Neoliberal Adjustment Policies and Local Initiatives,* edited by Liisa North and John Cameron. Bloomfield, CT: Kumarian Press.

Brown, David L. et al. (2000). 'Globalization, NGOs and Multisectorial Relations,' in Joseph S. Nye and John D. Donohue (eds.) *Governance in a Globalizing World.* Cambridge: Cambridge University Press, pp. 271–97.

Buchi, Hernán (1993). *La transformación económica de chile. Del estatismo a la libertad económica.* Bogotá: Editorial Norma.

Bulmer-Thomas, Victor (1996). *The New Economic Model in Latin America and its Impact on Income Distribution and Power.* New York: St. Martin's Press.

Burbach, Roger (1994). 'Roots of the Postmodern Rebellion in Chiapas,' *New Left Review,* 205.

Burki, S. and G. Perry (1998). *Más allá del consenso de Washington: la hora de la reforma institucional.* Washington, DC: World Bank.

Burnside, Craig and David Dollar (1997). *Aid, Policies and Growth.* Washington, DC: The World Bank.

CAAP (Centro Andino de Acción Popular) (1998). *Bases de datos.* http://www. ecuanex.net.ec.

Calderón, Fernando (1995). *Movimientos sociales y politica.* Mexico: Siglo XX1.

Calderón, Fernando and Norberto Lechner (1998). *Más allá del estado, más allá del Mercado: la democracia.* La Paz: Editorial Plural.

Cameron, John (2003). 'Municipal Democratization and Rural Development in Highland Ecuador,' in *Rural Progress, Rural Decay: Neoliberal Adjustment Policies and Local Initiatives,* edited by Liisa North and John Cameron. Bloomfield, CT: Kumarian Press.

Cardoso, Fernando Henrique (1995). 'Democracy and Development,' Address delivered on the occasion of his visit to ECLAC, Santiago, Chile, March 3.

Carothers, T. (1999). *Aiding Democracy Abroad.* Washington, DC: Carnegie Endowment for International Peace.

Carrasco, Hernán (1993). 'Democratización de los poderes locales y levantamiento indígena,' in *Sismo étnico en el Ecuador.* Quito: CEDIME— Abya-Yala, pp. 29–69.

Carrasco V. and Carlos Marx (1998). *Ecuador y el Consenso de Washington: la hora neoliberal.* Cuenca: University of Cuenca.

Carroll, Thomas (1992). *Intermediary NGOs. The Supporting Link in Grassroots Development.* West Hartford, CT: Kumarian Press.

CEDIB (2001). *Treinte Días de Noticias.* Cochabamba: CEDIB.

CEPAL (Comisión Económica para América Latina) (2000). *Equidad, desarrollo y ciudadanía.* México DF: CEPAL.

Chalmers, et al. (eds.) (1997). *The New Politics of Inequality in Latin America.* Oxford: Oxford University Press.

Chan, Yu Ping (2001). 'Democracy or Bust? The Development Dilemma,' *Harvard International Review,* Fall.

Cheema, S. and D. A. Rondinelli (eds.) (1983). *Decentralization and Development.* Beverly Hills: Sage.

Chiriboga, Manuel (1986). 'Crisis económica y movimiento campesino e indígena en Ecuador,' *Revista Andina,* 7, 4(4), July, Cusco, Centro Bartolomé de las Casas, pp. 7–30.

Chiriboga, Manuel (1987). 'Movimiento campesino y indígena y participación política en Ecuador: la construcción de identidades en una sociedad heterogénea,' *Ecuador Debate,* 13, May, Quito, CAAP, pp. 87–121.

CIPCA (Centro de investigación y promoción del Campesinado) (1992). *Futuro de la comunidad campesina.* La Paz.

CODENPE (Consejo de Nacionalidades y Pueblos del Ecuador) (1995). *Infocodenpe. Boletín Especial,* 28 Dic. Quito.

Coffey, Gerard (2004). 'Un paquetazo disfrazado,' *Tintaji,* 39.

Coffey, Ferardo (2002). 'La hora final del dollar,' *Quincenario Tintají* [Ecuador], November.

Coggiala, Osvaldo (2003). 'Del conflicto de enero a la revolución de octubre: parto y nacimiento de la revolución Boliviana,' *En Defensa del Marxismo,* 11(32), December.

Colectivo Situaciones (2001). *Contrapoder: una introducción.* Buenos Aires: Ediciones de Mano en Mano.

Collier, Paul (1997). 'The Failure of Conditionality,' in Catherine Gwin and Joan Nelson (eds.) *Perspectives on Aid and Development,* Washington, DC: ODI; Baltimore MD: Johns Hopkins University Press, pp. 31–77.

Comité Editorial (1986). 'Identidad, movimiento social y participación electoral,' *Ecuador Debate,* 12, December: 11–22.

CONAIE (Confederación de nacionalidades indígenas de Ecuador) (1994). *Proyecto político de la CONAIE.* Quito: CONAIE.

CONAIE (2003). *Mandato de la I Cumbre de las Nacionalidades, Pueblos y Autoridades Alternativas.* Quito: CONAIE.

CONAM (2000). *La democratización de los capitales en los procesos de privatización.* Quito: CONAIE.

Cornia, Andrea, Richard Jolly and Frances Stewart (eds.) (1987). *Adjustment with a Human Face.* Oxford: Oxford University Press.

Corragio, José Luis et al. (2001). *Empleo y economia del trabajo en el Ecuador. Algunas propuestas para superar la crisis.* Quito: ILDIS / Editorial Abya Yala.

Coyuntura / Tema Central (1996). 'Caras y mascaras del ajuste,' *Ecuador Debate,* 37, April.

Crozier, M., S. P. Huntington and J. Watanuki (1975). *The Crisis of Democracy: Report on the Governability of Democracies to the Trilateral Commission*. New York: New York University Press.

Dávilas, Pablo (2003). *Movimiento indígena ecuatoriano: la construcción de una utopia*. Manuscript, provided by author. Quito.

Dávilas, Pablo (2004). 'Las transformaciones políticas del movimiento indígena ecuatoriano,' *OSAL*, December. http://osal.clacso.org/espanol/html/frevista. html.

De la Fuente, Manuel (ed.) (2001). *Participación popular y desarrollo local*, Cochabamba: PROMEC-CEPLAG-CESU.

De Walt, Billie and Martha Ress (1994). *The End of Agrarian Reform in Mexico: Past Lessons and Future Prospects*, San Diego: Center for US-Mexican Studies.

Deacon, Bob (2000). 'Social Policy in a Global Context,' in Andrew Hurrell and Ngaire Woods (eds.) *Inequality, Globalisation and World Politics*. Oxford: Oxford University Press.

Deininger, K. (1998). 'Implementing Negotiated Land Reform: Initial Experience from Colombia, Brazil and South Africa,' in *Proceedings of the International Conference on Land Tenure in the Developing World with a Focus on Southern Africa*, January 27–29, University of Cape Town, pp. 116–25.

Delgadillo Terceros, Walter and Jonny Zambrana Barrios (2002). *Experiencias de los consejos de participación popular (CPPs)*. Cochabamba: PROSANA, Unidad de fortalecimiento comunitario y transversales.

Delgado Jara, Diego (2000). *Atraco bancario y dolarización*. Quito: Ediciones Gallo Rojo.

Deruytters, Anne (1997). *El BID y los pueblos indígenas*. Washington, DC: BID.

Diamond, Larry (1999). *Developing Democracy. Towards Consolidation*, Baltimore: Johns Hopkins University Press.

Dominguez, J. and A. Lowenthal (eds.) (1996). *Constructing Democratic Governance*, Baltimore: Johns Hopkins University Press.

Dror, Y (1994). *La capacidad de gobernar*. Barcelona: Circulo de Lectores.

Duran, B. J. (1990). *Las nuevas instituciones de la sociedad civil*. La Paz: Huellas S.R.L.

Echeverría, Julio (1997). *Democracia bloqueada: Teoría y crisis del sistema politico ecuatoriano*. Quito: Letras.

ECLAC (Economic Comission for Latin America and the Caribbean) (1990). *Productive Transformation with Equity*, Santiago: ECLAC.

Econoticias (2004a). 'El gobierno de Carlos Mesa.' http://www. econoticiasbolivia.com.

Econoticias (2004b). 'Las condiciones del Banco Mundial.' http://www. econoticias bolivia.com.

Ecuador, Presidencia de la Republica (2003). *Gestión del gobierno e instituciones del estado. enero-agosto 2003*. Quito.

ECUARUNARI (1998). *Historia de la nacionalidad y los pueblos quichuas del Ecuador*. Quito: Ecuador Runacupac Riccharimui.

Editorial (2000). 'Legitimidad y poder: Los limites de la práctica política actual,' *Boletín ICCI 'RIMAY,'* 3(21), December, pp. 1–4.

Edwards, Michael and David Hulme (1992). *Making a Difference: NGOs and Development in a Changing World.* London: Earthscan.

Edwards, Michael and David Hulme (1996). *Beyond the Magic Bullet: NGO Performance and Accountability in the Post-Cold War World.* Bloomfield, CT: Kumarian.

Equipo de Coyuntura (1998), 'Ecuador: la coyuntura de 1997/1998,' *Ecuador Debate,* No. 35, April.

Escobar, Arturo and Sonia Alvarez (eds.) (1992). *The Making of Social Movements in Latin America: Identity, Strategy, and Democracy.* Boulder, CO: Westview Press.

Finot, Iván (1997). *Descentralización y participación en América Latina: Como conciliar eficiencia con equidad.* Quito: ILPES.

Fisher, J. (1993). *The Road from Rio: Sustainable Development and the Nongovernmental Movement in the Third World.* Westport, CT: Praeger.

Fowler, Alan (1997). *Striking a Balance: A Guide to Enhancing the Effectiveness of NGOs in International Development.* London: Earthscan.

Freedom House (1999). *Freedom in the World: The Annual Survey of Political Rights and Civil Liberties 1998–1999.* New York: Freedom House.

Friedman, John (1992). *Empowerment: The Politics of Alternative Development.* Oxford: Blackwell.

Gannitsor, Irene (1998). 'Development Planning in Rural Bolivia: Limits and Constraints.' MA Thesis, University of Guelph.

Gannitsos, L. (1998). 'Popular Participation for Municipal Development in Rural Bolivia: Limits and Constraints.' Unpublished MA Thesis, University of Guelph.

Garcia Linera, Alvaro (2001). 'Indios y q'aras: la reinvención de las fronteras internas,' July. <http://www.clacso.edu.ar>

García Linera, Alvaro (2004). 'Democracia liberal versus democracia comunitaria,' *El Juguete Rabiosa* [La Paz], 3(96), 20 de Enero.

Graham, C. (1992). 'The Politics of Protecting the Poor during Adjustment: Bolivia's Social Emergency Fund,' *World Development,* 29(9): 1233–51.

Grindle, M.S. (2000). *Audacious Reforms: Institutional Invention and Democracy in Latin America.* Baltimore: Johns Hopkins University Press.

Gwin, Catherine and Joan M. Nelson (1997). *Perspectives on Aid and Development.* Washington, DC: Overseas Development Council.

Haggard, S. and R. Kaufman (1995). *The Political Economy of Democratic Transitions.* Princeton, NJ: Princeton University Press.

Harris, John (2001). *Depoliticising Development. The World Bank and Social Capital.* New Delhi: Left Word Books.

Hayden, Robert (2002). 'Dictatorships of Virtue,' *Harvard International Review,* Summer.

Heller, Henry (2003). *Canadian Dimension,* 37(6), Nov./Dec.

Helman, Judith (1995). 'The Riddle of New Social Movements: Who They Are and What They Do,' in S. Halebsky and R. Harris (eds.), *Capital, Power and Inequality in Latin America.* Boulder, CO: Westview Press.

Hernandez, Virgilio (1998). 'La historia del Jatun Oso,' in ECUARUNARI, *Historia de la nacionalidad y los pueblos quichuas del Ecuador.* Quito: ECUARUNARI, pp. 2111–213.

Holloway, John (2001). 'Doce tesis sobre el anti-poder,' in *Contrapoder: una introducción*, editada por Colectivo Situaciones. Buenos Aires: Ediciones de Mano en Mano (November), pp. 73–82.

Holloway, John (2002). *Cambiar el mundo sin tomar el poder*. Buenos Aires: Editor Andrés Alfredo Méndez.

Hunter, Allen (1995). 'Los nuevos movimientos sociales y la revolución,' *Nueva Sociedad*, 136: 3–4.

Huntington, S. P. (1991). *The Third Wave Democratization in the Late Twentieth Century*. Norman: University of Oklahoma Press.

Ibarra, Hernán (2002). 'Los movimientos étnicos y la redefinición de las relaciones Indígenas-Estado en Ecuador y México,' *Mimeo*, Abril. Quito: CAAP.

ICCI-RIMAY (2001). 'Banco Mundial y PRODEPINE: Hacia un neoliberalismo étnico?' Bolétin ICCI 'RIMAY,' No. 25, Abril.

IDB (Inter-American Development Bank) (2000). *Development beyond Economics. Economic and Social Progress in Latin America. 2000 Report*. Baltimore: Johns Hopkins University Press.

ILIS (Instituto Latinoamericano de Investigaciones Sociales) (2003). *Análisis de coyuntura económica del 2003*. Quito: ILDIS/Friedrich Ebert Stiftung.

ILPES/CAF (1996). *Marco regulatorio, privatización y modernización del Estado*. Santiago: ILPES/CAE.

INE (2002). *Mapa de la pobreza de Bolivia*. La Paz: INE.

INEC-Ecuador (2001). *Censo de población*. Quito: INEC.

INE (Instituto Nacional de Estadísticas) (2002). *Bolivia: Características de la población*. La Paz: INE.

Kamat, Sangeeta (2003). 'NGOs and the New Democracy: The False Saviors of International Development,' *Harvard International Review*, Spring.

Kapstein, Ethan (1996). 'Workers and the World Economy,' *Foreign Affairs*, 75(3).

Karl, T. L. (2000). 'Economic Inequality and Democratic Instability,' *Journal of Democracy*, XI (1): 149–56.

Kaufmann, Daniel, Art Kraay and Pablo Zoido-Lobatón (1999). *Governance Matters*. Washington, DC: World Bank.

Kliksberg, Bernardo (1999). 'Capital social y cultura, claves esenciales del desarrollo,' *Revista de la CEPAL*, 69, December.

Kliksberg, Bernardo (2001). 'Seis tesis no convencionales sobre participación,' in *Capital social y cultura: claves estrategícos para el desarrollo*, edited by Bernardo Kliksberg and Luciano Tomasini. Buenos Aires: BID-Fondo Cultural Económico.

Knack, S. (1999), 'Social Capital, Growth and Poverty: A Survey of Cross-Country Evidence,' *Social Capital Initiative Working Paper 7*, World Bank, Social Development Department, Washington, DC.

Kohl, Benjamin (1999a). 'Economic and Political Restructuring in Bolivia: Tools for a Neoliberal Agenda?' Unpublished PhD Dissertation, Cornell University.

Kohl, Benjamin (1999b). 'The Role of NGOs in Implementing Political and Administrative Decentralization in Bolivia,' Paper presented at the American Colleges and Schools of Planning conference, Chicago, October 23.

Kohl, Benjamin (2002). 'Stabilizing Neoliberalism in Bolivia: Popular Participation and Privatization,' *Political Geography*, 21: 449–72.

Kohli, A. (1995). *Democracy and Discontent*. Cambridge: Cambridge University Press.

Krueger, Anne (1974). 'The Political Economy of the Rent-Seeking Society,' *American Economic Review*, 64(3).

Krueger, Anne, C. Michalopoulos and V. Ruttan (1989). *Aid and Development*. Baltimore: Johns Hopkins University Press.

Kruse, Tom (2002). 'La segunda batalla en la guerra del agua,' CEDIB, La Paz. August. tkruse@albatross.cnb.net.

Kumar, K. (1993). 'Civil Society: an Inquiry into the Usefulness of an Historical Term,' *British Journal of Sociology*, 44(3), pp. 375–401.

Larrea, Carlos (1999). *Desarrollo social y gestión municipal en el Ecuador: Jerarquización y tipología*. Quito: ODEPLAN.

Larrea, Carlos (2004). *Pobreza, dolarización y crisis en el Ecuador*. Quito: Editorial Abya-Yala.

Larrea, Carlos and Liisa North (1997). 'Ecuador: Adjustment Policy Impacts on Truncated Development and Democratization,' *Third World Quarterly*, 18(5): 913–34.

Latham, Robert (1997). 'Globalisation and Democratic Provisionism: Re-reading Polanyi,' *New Political Economy*, 2(1).

Lewis, David and Tina Wallace (eds.), (2003). *New Roles and Relevance: Development NGOs and the Challenge of Change*. West Hartford, CT: Kumarian.

Lindenberg, Mark and Coralie Bryant (2002). *Going Global: Tranforming Relief and Development NGOs*. Bloomfield, CT: Kumarian.

Litvack, Jennie, Junaid Ahmad and Richard Bird (1998). 'Rethinking Decentralization in Developing Countries,' *World Development Sources*, Sector Studies Series. Washington DC: World Bank.

Lluco Tixe, Miguel (1998). 'El levantamiento indígena dew 1990, o la mecha que prendio el pajonal,' in Ecuarunari, *Historia de la nacionalidad y los pueblos quichuas del Ecuador*. Quito: ECUARUNARI, pp. 199–210.

Londoño, J. L. (1996). *Pobreza, desigualdad y formación del capital humano en América Latina*. Washington, DC: World Bank.

Lowenthal, A. F. (1999). 'Latin America at the Century's Turn,' *Journal of Democracy*, XI (2): 412–55.

Lucas, Kintto (2003). *El movimiento indígena y las acrobacias del coronel*. Quito: Libros deTintaji.

Macas, Luis (2000). 'Movimiento indígena ecuatoriano: Una evaluación necesaria,' *Boletín ICCI 'RIMAY,'* 3(21), December, pp. 1–5.

Macas, Luis (2004). 'El movimiento Indígena: Aproximaciones a la comprensión del desarrollo ideológico politico,' *Tendencia Revista Ideológico Político*, I, Quito, March, pp. 60–7.

Macdonald, Laura (1997). *Supporting Civil Society: The Political Role of NGOs in Central America*, London: Macmillan Press.

MAGDR (Ministerio de Agricultura, Ganadería y Desarrollo Rural) (1999). *Municipio Productivo: Promoción Económica Rural*. La Paz: MAGDR.

Maizels, A. and M.K. Mssanke (1984). 'Motivations for Aid to Developing Countries,' *World Development*, September, 100: 879–901.

Mallimaci, F. (1996). 'Políticas sociales: hacia una nueva relación entre estado sociedad civil. Las organizaciones no gubermentales de promoción y desarrollo,' *Dialógica* No. 1. Buenos Aires: CEIL.

Mamani Ramirez, Pablo (2003). 'El rugir de la multitud: levantamiento de la ciudad aymara de El Alto y caída del gobierno de Sánchez de Lozada,' *OSAL (Obervatorio Social de América Latina)*, IV(12), Sep./Dec.

Martinez Nogueira, Roberto (1991). 'Los pequeños proyectos: ¿microsoluciones para macroproblemas?' in Roberto Martinez Nogueira (ed.) *La trama solidaria.pobreza y microproyectos de desarrollo social.* Buenos Aires: GADIS, Imago-Mundi.

Mawdsley, Emma, Janet Townsend, Gina Porter and Peter Oakley (2002). *Knowledge, Power and Development Agendas: NGOs North and South.* Oxford: Intrac Publications.

Mayorga, Fernando, ed. (1997). *¿Ejemonias? Democracia representiva y liderazgos locales.* La Paz: PIEB.

Mayorga, René (1997). 'Bolivia's Silent Revolution,' *Journal of Democracy*, 8(1): 142–56.

McNeish, John (2003). 'Globalization and the Reinvention of Andean Tradition: The Politics of Community and Ethnicity in Highland Bolivia,' in *Latin American Peasants*, edited by Tom Brass. London: Frank Cass, pp. 228–69.

MCS-Ministerio de Vivienda y Servicios Básicos (2002). *Guía de desarrollo comunitario para proyectos de agua y saneamiento.* La Paz.

MDH-SNPP (Ministerio de Desarrollo Humano-Secretaria Nacional de Participación Popular) (1997). *Bolivia: La participación popular en cifras*, II. La Paz.

Medina, Javier (2001). *Manifiesto Municipalista: por una democracia participative.* La Paz, G-DRU Grupo Interinstitutional de Desarrollo Rural.

Melucci, Alberto (1992). 'Liberation or Meaning: Social Movements, Culture, and Democracy,' in J. Nederveen Pieterse (ed.) *Emancipations, Modern and Postmodern.* London: Sage.

Messner, Dirk (2001). 'Globalización y gobernabilidad global,' *Nueva Sociedad*, Nov./Dec., 176.

Miró, Joseph (1998). 'The Law of Popular Participation: A Neoliberal Bolivian Opening.' Unpublished manuscript.

Mitlin, Diana (1998). 'The NGO Sector and its Role in Strengthening Civil Society and Securing Good Governance,' in Armanda Bernard, Henry Helmich and Percy Lehning (eds.), *Civil Society and International Development.* Paris: OECD Development Center.

MNR-MRTKL (Movimiento Nacional Revolucionario-Movimiento Revolucionario Tupac Katari) (1993). *El Plan de Todos.* La Paz MNR-MRTKL.

Molina, M. Fernando (1997). *Historia de la participación popular.* La Paz: Ministerio de Desarrollo Humano.

Montúfar, César (n.d.). *La reconstrucción neoliberal: Febres Cordero o la estatización del neoliberalismo en el Ecuador 1984–1988.* Quito: Ediciones Abya-Yala.

Morales, Evo (2003). 'La hoja de coca, una bandera de lucha,' Interview with *Punto Final* (Santiago), May.

Morton, Adam (2004). 'The Antiglobalization Movement: Juggernaut or Jalopy?,' in Henry Veltmeyer (ed.), *Globalization and Antiglobalization: Dynamics of Social Change in the New World Order*. London: Ashgate, pp. 155–68.

Mosley, Paul (1999). 'Globalization, Economic Policy and Growth Performance,' *International Monetary and Financial Issues for the 1990s*, X. New York and Geneva: United Nations, pp. 157–74.

Muñoz, J. Francisco (ed.) (1999). *Descentralización*. Quito: Editorial Tramasocial.

Navarro Jiménez, Guíllermo (2000). *Ecuador. Corrupción, política económica y gobernabilidad*. Quito.

Navarro Jiménez, Guíllermo (n.d.). *Capitalismo popular, privatizaciones y concentración económica*. Quito: Ediciones Zitra.

Negri, Antonio (2001). 'Contrapoder,' in *Contrapoder: una introducción*. Buenos Aires: Ediciones de Mano en Mano, pp. 83–92.

OECD (1997). *Final Report of the DAC Ad Hoc Working Group on Participatory Development and Good Governance*. Paris: OECD.

Okonski, Kendra (2001). 'Riots Inc. The Business of Protesting Globalization,' *The Wall Street Journal*, editorial page, August 14.

OrruñoYañez, Armando (2002). 'Hacia una nueva articulación de las políticas sobre la pobreza en Bolivia,' in *Bolivia: visiones de futuro*. La Paz: FES-ILDIS, pp. 209–52.

Ottaway, Marina (2003). *Democracy Challenged: The Rise of Semi-Authoritarianism*. Washington, DC: Carnegie Endowment for International Peace.

Oxhorn, Philippe (1999). 'Construction of the State by Civil Society: Bolivia's Law of Popular Participation and the Challenge of Local Democracy,' McGill University.

PADEM (Programa de Apoyo a la Democracia Municipal) (2002). *Empoderamiento de las comunidades campesinas y indígenas: una propuesta para la democratización de los municipios rurales*. <http://www.servicioweb. cl/ indice_web/Comunitarios.htm>.

Palma Carbajal, Eduardo (1995). 'Decentralization and Democracy: The New Latin American Municipality,' *CEPAL Review*, 55, April.

Payne, M. (1999), 'Instituciones política e instituciones económicas: Nueva vision sobre las relaciones entre el estado y el Mercado,' *Reforma y Democracia*, XIII(119):140.

Payne, M., Daniel Zovatto, Fernando Carillo Flórez and Andres Allemand Zavala (2003). *La política importa. Democracia y desarrollo en América Latina*. Washington, DC: BID.

Peirce, Margaret (ed.) (1997). *Capitalization: The Bolivian Model of Social and Economic Reform*. Miami: Woodrow Wilson Center and the North South Center.

Petras, James and Henry Veltmeyer (2001). *Unmasking Globalization: The New Face of Imperialism*. London: Zed Books; Halifax: Fernwood Books.

Petras, James and Henry Veltmeyer (2003). 'The Peasantry and the State in Latin America,' in *Latin American Peasants*, edited by Tom Brass. London: Frank Cass, pp. 41–82.

PNUD (Programa de las Naciones Unidas del Desarrollo) (1999). *Informe sobre Desarrollo Humano. Ecuador 1999.* Quito: PNUD. http://www.transparency. org/tilac/indices/indices percepcion/2002/ipc2002.html.

PNUD (Programa de Naciones Unidos de Desarrollo) (2002). *Informe de Desarrollo Humano.* La Paz: PNUD.

Przeworski, Adam (1995). *Democracia Sustentable.* Buenos Aires: Editorial Paidós SAICF.

Przeworski, Adam (2003). 'A Flawed Blueprint: The Covert Politicization of Development Economics,' *Harvard International Review,* Spring.

Quispe Huanca, Felipe (1999). *Tupak Katari vive y vuelve ... carajo.* Oruro: Quelco.

Reilly, Charles (1989). *The Democratization of Development: Partnership at the Grassroots.* Arlington: Inter-American Foundation Annual Report.

Retolaza Eguren, Iñigo (2003). 'El municipio somos todos. Gobernancia participativa y transferencia municipal,' *Medicus Mundi.* La Paz: Ed. Plural.

Rhon Dávila, Francisco (2003). 'Estado y movimientos étnicos en Ecuador,' in *Movimientos sociales y conflicto en América Latina,* edited by José Seoane. Buenos Aires: CLACSO, pp. 127–40.

RIAD (Red Interamericana Agricultura y democracia) (1998). *Organizaciones campesinas e indígenas y poderes locales: propuestas para la gestión participativa del desarrollo local.* Quito: RIAD/ Grupo Democracia y Desarrollo Local/ Editorial Abya-Yala.

Riddell, Roger and Mark Robinson (1997). *NGOs and Rural Poverty Alleviation.* London: Overseas Development Institute.

Rodríguez, Ostra (1995). *Estado y municipio en Bolivia.* La Paz: SNPP and PNUD.

Rodrik, Dani (1995). 'Why is there Multilateral Lending?' *Annual World Bank Conference on Development Economics,* edited by Michael Bruno and Boris Pleskovic. Washington, DC: The World Bank.

Rondinelli, D. A. (1989). 'Implementing Decentralization Programs in Asia: A Comparative Analysis,' *Public Administration and Development,* 3 (3): 181–207.

Rondinelli, D. A., J. McCullough and W. Johnson (1989). 'Analyzing Decentralization Policies in Developing Countries: A Political Economy Framework,' *Development and Change,* 20(1): 57–87.

Rondinelli, D. A., J. R. Nellis and G. S. Cheema (1983). 'Decentralization in Developing Countries: A Review of Recent Experience,' *World Bank Staff Paper,* No. 581. Washington, DC: World Bank.

Rueschmeyer, D. and E. H. Stephens (1992). *Capitalist Development and Democracy.* Chicago: University of Chicago Press.

Salbuchi, Adrian (2000). *El cerebro del mundo: la cara oculta de la globalización.* Córdoba: Ediciones del Copista.

Salgado, Wilma (2000). 'Recuperación a pesar de la dolarización y el ajuste,' *Ecuador Debate,* No. 50 (CAAP, Quito).

Salop, Joanne (1992). 'Reducing Poverty: Spreading the Word,' *Finance & Development,* 29(4), December.

Saltos Galarza, Napoleón (2001). 'Movimiento indígena y movimientos sociales: Encuentros y desencuentros,' *Bolétin ICCI RIMAY,* 3(27), June.

Salvatierra, Hugo (1999). 'Indigenous Strategy: Organic Community Strengthening, Self-Management of Territories and Natural Resources, and Municipal Administration.' Unpublished manuscript. Santa Cruz.

Sánchez-Parga, José (ed.) (1993) *Etnia, poder y diferencia en los Andes septentrionales.* Quito: Abya Yala.

Sánchez, Parga, José (1998). 'La cultura entre un fin de un siglo y umbral de un milenio,' mimeo. Quito: CAAP.

Sánchez, Rolando (ed.) (2003). *Desarrollo pensado desde los municipios: capital social y despliegue de potencialidades local.* La Paz: PIED (Programa de Investigación Estratégia en Bolivia).

Sánchez, Rolando (ed.) (2003). *Desarrollo pensado desde los municipios: capital social y despliegue de potencialidades local.* La Paz: PIED (Programa de Investigación Estratégia en Bolivia).

Santana, Roberto (1984). *El campesinado indígena y el desafío de la modernidad.* Quito: CAAP.

SAPRIN Ecuador, Red de la Sociedad Civil para las Alternativas Económicas (2004). *Los impactos del neoliberalismo: Una lectura distinta desde la percepción y experiencias de los actores.* Quito: Ediciones Abya Yala.

Schuller, Tom et al. (eds.) (2000). *Social Capital: Critical Perspectives.* Oxford University Press.

Sen, Amartya (1999). *Development as Freedom.* New York: Alfred Knopf.

SIOS (Sistema de Información de Organizaciones Sociales) (2002). *Directorio. Organizaciones Sociales de Desarrollo.* Quito Fundación Alternativa.

Slater, David (1985). *New Social Movements and the State in Latin America.* Amsterdam: CEDLA.

Solón, Pablo (2004). 'Conjuntura de Bolivia: bela entrevista com o intellectual Pablo Solón,' Interview May 31. astedile@uol.com.br.

Stein, Alfredo (1991). 'Las ONGs y su rol en el desarrollo social de América Latina,' in *La encrucijada de los 90. América Latina en Pensamiento Iberoamericano.* Madrid.

Stiefel, Matthias and Marshall Wolfe (1994). *A Voice for the Excluded: Popular Participation in Development: Utopia or Necessity?* London and Atlantic Highlands, NJ: Zed Books and UNRISD.

Stiglitz, Joseph. (2002). *Globalization and Its Discontents.* New York: W. W. Norton.

Sulbrandt, J. (1994). 'Presidencia y gobernabilidad en América Latina: de la presidencia autocrática a la democrática,' *Reforma y Democracia*, 2, Caracas.

Tendencia—Revista Ideológico Político (2004). *Tema Central: Descentralización.* 1 March.

Tendler, Judith (1975). *Inside Foreign Aid.* Baltimore: Johns Hopkins University Press.

Thompson, A. (1995). *Público y privado. Las organizaciones sin fines de lucro en la Argentina.* Buenos Aires: UNICEF / LOSADA.

Thompson, A. (1990). 'El tercer sector y el desarrollo social,' in *Mucho, poquito y nada.* Buenos Aires: UNICEF.

Torres D., Victor (n.d.). *Sistema de desarrollo local, Sisdel.* Quito: Ediciones Abya-Yala.

Toye, John (1987). *Dilemmas of Development: Reflections on the Counter-Revolution in Development Theory and Policy*. Oxford: Basil Blackwell.

UNDP (1996). 'Good Governance and Sustainable Human Development,' *Governance Policy Paper*. http://magnet.undp.org/policy.

UNRISD (2000). 'Civil Society Strategies and Movements for Rural Asset Redistribution and Improved Livelihoods,' UNRISD—Civil Society and Social Movements Program. Geneva: UNRISD.

Utting, Peter (2000). 'UN–Business Partnerships: Whose Agenda Counts?' *UNRISD News*, 23, Autumn–Winter.

Van Cott, D. L. (2000). *The Friendly Liquidation of the Past: The Politics of Diversity in Latin America*. Pittsburgh: University of Pittsburgh Press.

Vargas, Humberto and Eduardo Córdova (2003). 'Bolivia: un país de re-configuraciones por una cultura de pactos politicos y de conflictos,' *Movimientes sociales y conflicto en América Latina*. Buenos Aires: CLACSO.

Vásquez, Lola and Napoleón Saltos (2003). *Ecuador: su realidad*. Quito: Fundación José Peralta.

Veltmeyer, Henry (1992). 'Social Exclusion and Rural Development in Latin America,' *Canadian Journal of Latin American and Caribbean Studies*, 27(54).

Veltmeyer, Henry (1997). 'Decentralisation as the Institutional Basis for Participatory Development: The Latin American Perspective,' *Canadian Journal of Development Studies*, XVIII(2).

Veltmeyer, Henry (2002). 'Social Exclusion and Rural Development in Latin America,' *Canadian Journal of Latin American and Caribbean Studies*, 27(54).

Veltmeyer, Henry and Anthony O'Malley (2001). *Transcending Neoliberalism: Community-Based Development*. West Hartford, CT: Kumarian Press.

Veltmeyer, Henry and James Petras (1997). *Economic Liberalism and Class Conflict in Latin America*. London: MacMillan.

Veltmeyer, Henry and James Petras (2000). *The Dynamics of Social Change in Latin America*. London: Macmillan.

Vernon, Raymond (1992). *La promesa de la privatización. Un desafío para la política exterior de los Estados Unidos*. México: Fondo de Cultura Económica.

Vicuña Izquierdo, Leonardo (2000). *Política económica del Ecuador: Dos décadas perdidas. Los años 80–90*. Guayaquil: ESPOL.

Villegas Quiroga, Carlos (2003). 'Rebelión popular y los derechos de propiedad de los hidrocarburos,' *OSAL—Obervatorio Social de Améríca Latina*, IV(12), Sept./Dec.

Viteri, Hugo Galo (1998). *Las políticas de ajuste en Ecuador, 1982- 1996*. Quito: Corporación Editora Nacional.

VMPPFM (1998). Avances, limitaciones y desafíos de la planificación participativa municipal. http://www.vppfm.gov.bo/vppfm/PLANIF_PAR/avandesa.htm.

Wallace, Tina (2003). 'NGO Dilemmas: Trojan Horses for Global Neoliberalism?' *Socialist Register 2004*. London: Merlin Press.

Wallace, Tina, Sarah Crowther and Andrew Shephard (1997). *Standardising Development: Influences on UK NGOs Policies and Procedures*. Oxford: Westview Press.

Weber, Heloise (2002). 'Global Governance and Poverty Reduction: the Case of Microcredit,' in Rorden Wilkinson and Steve Hughes (eds.) *Global Governance: Critical Perspectives*. London and New York: Routledge, pp. 132–51.

Werlin, H. (1992). 'Linking Decentralization and Centralization: A Critique of the New Development Administration,' *Public Administration and Development*, 12 (3): 223–35.

White, Howard (1994). 'The Countrywide Effects of Aid,' *World Bank Policy Research Working Paper*, No. 1334. Washington, DC: The World Bank.

Williamson, John (ed.) (1990), *Latin American Adjustment. How Much Has Happened?* Washington, DC: Institute for International Economics.

Woolcock M. and D. Narayan (2000). 'Social Capital: Implications for Development Theory, Research and Policy,' *The World Bank Research Observer*, 15(2), August.

World Bank (1989). *Sub-Saharan Africa: from Crisis to Sustainable Growth*. Washington, DC: World Bank.

World Bank (1994). *Governance. The World Bank Experience*. Washington, DC: World Bank.

World Bank (1995). *Workers in an Integrating World*. New York: Oxford University Press.

World Bank (1996). *Ecuador Poverty Report*. Washington, DC: World Bank.

World Bank (1997). *The State in a Changing World: World Development Report*. Oxford: Oxford University Press.

World Bank (1998). *Assessing Aid. What Works, What Doesn't, and Why*. New York: Oxford University Press.

Xue, lan (1997). The Capitalization Program in Bolivia and its Implications for State-Owned Enterprise (SOE) Reform in China,' in Margaret Peirce (ed.) *Capitalization: The Bolivian Model of Social and Economic Reform*. Miami: Woodrow Wilson Center and the North-South Center, pp. 239–57.

Zakaria, F. (1998). *El surgimiento de la democracia iliberal*. Quito: BID, Programa de Apoyo al Sistema de Gobernabilidad Democrática.

Zamosc, Léon (1993). Protesta agraria y movimiento indígena en la sierra ecuatoriana,' in *Sismo étnico en el Ecuador*. Quito: CEDIME—Abya Yala, pp. 273–304.

Index

Compiled by Mark Rushton